The Fascination with Unknown Time

Sibylle Baumbach · Lena Henningsen
Klaus Oschema
Editors

The Fascination with Unknown Time

Editors
Sibylle Baumbach
Department of English
University of Innsbruck
Innsbruck, Tirol
Austria

Klaus Oschema
Department of History
Ruhr University Bochum
Bochum, Nordrhein-Westfalen
Germany

Lena Henningsen
Department of Chinese Studies
University of Freiburg
Freiburg, Baden-Württemberg
Germany

ISBN 978-3-319-66437-8   ISBN 978-3-319-66438-5   (eBook)
DOI 10.1007/978-3-319-66438-5

Library of Congress Control Number: 2017950717

Cover design by Henry Petrides

Printed on acid-free paper

This Palgrave Macmillan imprint is published by Springer Nature
The registered company is Springer International Publishing AG
The registered company address is: Gewerbestrasse 11, 6330 Cham, Switzerland

Preface

The subject of this volume is so vast and rich that it can only be explored in broad interdisciplinary dialogue. Our aim was to include a wide range of disciplines from the humanities to address different facets of the fascination with time and to open up new avenues for its analysis by transgressing established disciplinary boundaries. This endeavor was largely made possible by the fertile and flexible framework for transdisciplinary collaboration offered by the German Young Academy (*Die Junge Akademie*) at the Berlin-Brandenburg Academy of Sciences and Humanities and the German National Academy of Sciences Leopoldina. Since its foundation in 2000, *Die Junge Akademie* has enabled vibrant cooperation and exchange among researchers from a wide range of disciplines in the sciences and the humanities. It also facilitated a project in which three scholars, working on early modern English literature, contemporary Chinese Studies, and medieval history, respectively, collaborated to explore a challenging and multi-faceted subject. Our work in the context of the Young Academy's Research Group "Fascination" resulted, amongst others, in a three-day conference at Berlin (9–11 July 2014) at which the first drafts of the contributions to this volume were presented and discussed in a committed, vivid, and—most of all—friendly and productive atmosphere. Based on this conference and the great potential of the transdisciplinary approach to the fascination with time, we decided to compile a volume on the topic to introduce this new research area and promote further research on the topic.

We thank all the contributors for engaging in the interdisciplinary dialogue, both at the initial conference and during the preparation of this volume, and for their constructive feedback and their patience during the editing process. We are particularly grateful to Walther Ch. Zimmerli, who, as our keynote speaker at the conference but also beyond, provided important input from a philosophical perspective. Our thanks also go to the Young Academy for funding a conference on this topic and to the team of the Young Academy's office in Berlin who helped tremendously with the logistics connected with both the conference and numerous meetings of the working group on "Fascination." We also thank Christof Diem for his editorial assistance in preparing this volume for publication. Finally, we thank the anonymous reviewers for their helpful suggestions and the editors at Palgrave Macmillan for their superb support in bringing this book to publication.

Innsbruck, Austria   Sibylle Baumbach
Freiburg, Germany   Lena Henningsen
Bochum, Germany   Klaus Oschema

Acknowledgements

The preparation of this publication has been generously supported by *Die Junge Akademie*.

The *Junge Akademie* was founded in 2000 as the first academy for the new academic generation worldwide. The fifty members of the *Junge Akademie*, young academics and artists from German-speaking countries, are dedicated to interdisciplinary discourse and are active at the interfaces between academia and society. The *Junge Akademie* is supported by the Berlin-Brandenburg Academy of Sciences and Humanities (BBAW) and the German National Academy of Sciences Leopoldina. The office is located in Berlin.

Contents

EDITORS AND CONTRIBUTORS

About the Editors

Sibylle Baumbach (Ph.D., Ludwig-Maximilians-University, Munich; Habilitation, Justus-Liebig-University, Giessen) is Professor of English Literature at the University of Innsbruck. Between 2011 and 2016, she was a member of the German Young Academy. Her research interests focus on early modern literature and culture, Shakespearean drama, literary and cultural theory, the aesthetics of fascination, and literary attention. Her publications include monographs on *Literature and Fascination* (2015), *Shakespeare and the Art of Physiognomy* (2008), and (co-)edited volumes on *Regions of Cultures–Regions of Identity* (2010), *A History of British Drama* (2011), *Travelling Concepts, Metaphors and Narratives* (2012), and *A History of British Poetry* (2015).

Lena Henningsen holds a Ph.D. in Chinese Studies from Heidelberg University. She currently is a member of the German Young Academy and a junior professor at the Institute of Chinese Studies at the University of Freiburg. Her research focuses on popular Chinese literature from the twentieth and twenty-first centuries, on practices of literary creation, circulation, and consumption, and on questions of authenticity and authorship. Her current research project is concerned with illegal entertainment fiction from and reading practices during the Chinese Cultural Revolution.

Klaus Oschema is professor for Late Medieval History at the Ruhr-Universität Bochum. He received his Ph.D. from TU Dresden and the EPHE Paris (2004) and his habilitation from Heidelberg University (2012). Between 2009 and 2014 he was a member of the German Young Academy, in 2016–2017 he was Gerda Henkel Member of the Institute for Advanced Study at Princeton, NJ (USA). His research focuses on late medieval court culture and nobility (especially Burgundy and France), medieval concepts of socialisation and their performative counterparts, and large-scale concepts of world order in the Middle Ages. He currently investigates late medieval astrologers as experts and scientific political advisers. Recent publications include *Bilder von Europa im Mittelalter* (2013) and *Die Performanz der Mächtigen* (co-editor, 2015).

Contributors

Daniel E. Agbiboa is Assistant Professor of Conflict Analysis and Resolution at George Mason University's School of Conflict Analysis and Resolution (S-CAR), United States. Before joining S-CAR, he completed a Postdoctoral Fellowship at the University of Pennsylvania's Perry World House, researching 'Global Shifts: Urbanization, Migration and Demography.' He holds a D.Phil. in International Development from the University of Oxford (St. Antony's College), and an M.Phil. in Development Studies from the University of Cambridge (Magdalene College). He has published extensively on conflict dynamics and urban politics in Africa in journals such as *African Affairs, Third World Quarterly,* and the *International Journal of Urban and Regional Research.*

Christian Hoffarth is a research and teaching fellow at the University of Duisburg-Essen, Germany. He holds a Ph.D. in Medieval and Modern History from the University of Hamburg (2016) and an M.A. from Heidelberg University (2010). From 2012 to 2014 he received a doctoral fellowship from Gerda Henkel Stiftung, Düsseldorf. In 2016 he was an invited visiting scholar at McGill University, Montreal. Hoffarth is author of *Urkirche als Utopie* (Franz Steiner, 2016) and has presented papers and published articles on medieval religious and social thought, biblical exegesis, the mendicant orders, and the Reformation reception of medieval heterodoxy.

Anke Holdenried is a Senior Lecturer in Medieval History (University of Bristol, UK). She works on the cultural and intellectual history of the Middle Ages and is particularly interested in medieval prophetic traditions. Holdenried has published on medieval political propaganda in the form of apocalyptic prophecy. Central to her current research is the broader network of medieval ideas about prophecy, including (but not limited to) its origin, its authority, its content, its relationship with time, its utility, and its limits.

Dorothee Xiaolong Hou is a Ph.D. student in Comparative Literature at the University of California, Davis, USA. Her current research and teaching interests include twentieth century and contemporary Chinese literature, cinema and visual culture, with a special interest in the construction of urban spaces and the shaping of everyday life.

András Kraft is a Ph.D. candidate in Medieval Studies at the Central European University (Budapest), where he specializes in the intellectual history of the Middle Byzantine period. His research focuses on how Byzantine philosophical and prophetic literature treats aspects of time. He holds an M.A. degree in Philosophy (Eötvös Loránd University, 2010) and an M.A. in Medieval Studies (Central European University, 2011). In 2014–2015 he was a junior fellow at the Research Center for Anatolian Civilizations (Istanbul), and he carried out research in Ioannina and Thessaloniki as a junior fellow of the Alexander S. Onassis Foundation in 2013–2014 and 2015–2016.

Hauke Lehmann is a film scholar. He is a postdoctoral researcher in a subproject of the Collaborative Research Center *Affective Societies* at the Freie Universität Berlin: "Migrant Melodramas and Culture Clash Comedies: Modulating a Turkish-German Sense of Commonality." His Ph.D. thesis will appear in 2017 as *Splitting the Spectator. An Affective History of the New Hollywood*. This book conceptualizes film history as a history of affective experience, based on the interplay of different cinematic modes of affectivity. Lehmann has published on cinematic texture, on figurations of the social, on psychedelic cinema, and on migration in film.

Sheldon H. Lu is Professor and Chair of Comparative Literature at the University of California, Davis, USA. He is the author and editor of ten books on Chinese literature, Chinese cinema, and comparative literature, including *Chinese Modernity and Global Biopolitics: Studies in Literature* *and Visual Culture* (University of Hawaii Press, 2007) and *China, Transnational Visuality, Global Postmodernity* (Stanford University Press, 2001).

Anna G. Piotrowska is mainly interested in researching sociological and cultural aspects of musical life and its role in determining political processes. She is the author of several books (including *Gypsy Music in European Culture*, 2013) and numerous scholarly articles. Currently associated with the Institute of Musicology at Jagiellonian University in Kraków (Poland), she has held many fellowships and awards as well as participated in several scientific projects, conferences, and workshops.

Caroline Rothauge is an Assistant Professor for Modern and Contemporary History at the Catholic University Eichstätt-Ingolstadt (Germany). Her dissertation considered questions of collective remembrance in Spanish audiovisual media, and her postdoctoral thesis focuses on temporal assumptions and practices in people's everyday lives in the German Kaiserreich around 1900. Currently, her research interests thus mainly relate to "time" as an object of historical investigation, but also include projects on the reciprocal effects between memory, history, and foreign affairs and on historical representations in so-called "quality TV."

Marco Tamborini holds a Ph.D. in History and Philosophy of Science from Heidelberg University. He is currently a research assistant at the Institute of Philosophy at the Technical University Darmstadt. His research focuses on the history and philosophy of biology in the nineteenth and twentieth centuries. In addition to publishing several articles on the history of biology, museums, and philosophy of science, he is currently working on a book manuscript on the visual and quantitative language of early nineteenth-century cameralism and natural history.

Katja Wehde studied German, English, and Education at the TU Dresden, complemented by cultural and media studies during a DAAD-funded stay at the English and Foreign Languages University in Hyderabad. In 2014, she received an M.A. in Transcultural Studies with a focus on Visual and Material Culture Studies from Heidelberg University. She has been awarded a doctoral scholarship from the International Graduate Centre for the Study of Culture at the Justus-Liebig-University in Giessen, where she is currently working on her dissertation project on the communication of cultural otherness in guided tours at German and British museums.

Kai Wiegandt studied English and German literature and philosophy at Universität Freiburg, Yale University, and at Freie Universität Berlin, where he is assistant professor of English. He is the author of *Crowd and Rumour in Shakespeare* (2012) and has published widely on early modern, modernist, and postcolonial literature and literary theory. In 2016, he received his habilitation at the Freie Universität Berlin with a study of J.M. Coetzee's posthumanism (forthcoming). He is an elected member of the German Young Academy.

LIST OF FIGURES

Time in the Making: Why All the Fuss About Time? On Time, the Unknown, and Fascination

Sibylle Baumbach, Lena Henningsen and Klaus Oschema

TIME AND THE UNKNOWN

In modern, Western societies, the effects of acceleration and an accompanying general 'lack of time' seem to be ubiquitous phenomena. They are, however, by no means particularly characteristic for the immediate past and the increasingly widespread use of digital devices. In fact, as early as 1973, German writer Michael Ende first published his novel *Momo*, subtitled "The strange story of the time-thieves and the child who

S. Baumbach (✉)
Department of English, University of Innsbruck, Innsbruck, Austria

L. Henningsen
Institute for Chinese Studies, University of Freiburg,
Freiburg, Germany

K. Oschema
Department of History, Ruhr-Universität Bochum, Bochum, Germany

© The Author(s) 2017
S. Baumbach et al. (eds.), *The Fascination with Unknown Time*,
DOI 10.1007/978-3-319-66438-5_1

1

brought the stolen time back to the people." Many readers of this fantasy novel or viewers of the successful movie that was based on it (1986) will remember the curious figures of the "Men in Grey" who represent the "Timesavings Bank." In keeping with modern ideals of efficiency and an economised perception of time, the Men in Grey promise their clients to help them save time, which can then be deposited with interest at their bank. As things turn out, this promise is deceiving, because the time people believe to have saved actually is lost (Ende 1973).[1]

Even though this insight has all the ingredients necessary for a good pun, it cannot explain the overwhelming success of Ende's novel, which went into numerous editions and also inspired several movies, audiobooks, and even musical works, including an opera. There must be more to it—something that explains the intense attraction, even fascination, which the subject of time exerts on readers and audiences. Several facets of this fascination with time are explored in the contributions to the present volume. As argued in these essays, one crucial aspect contributing to the fascination with time consists in the ambivalence that characterises modern perceptions of—and attitudes towards—time. This point can be confirmed by a brief glimpse at contemporary discussions in the media: although one discursive strand argues for the more efficient, even economic use of time (things have to run faster, etc.), there is also an unruly countercurrent that underlines the dangers of this development, stressing the need to use time in an integral and meaningful way. The parallel existence of both currents demonstrates the difficulties in arriving at a unanimous attitude towards time. But the problems do not begin only when judgments of value come into play: in fact, they are much more fundamental and start with the attempt to define 'time.'

"I know well enough what [time] is, provided that nobody asks me; but if I am asked what it is and try to explain, I am baffled" (Saint Augustine, *Confessions*, trans. Pine-Coffin, 1961: 17).[2] As implied by Augustine in his *Confessions*, time ultimately eludes our understanding. Even though we tend to refer to time as a universal phenomenon that

---

[1] For a critical analysis of the underlying ideas on time and their potential relationship to Buddhist perspectives, see Goodhew and Loy (2002).

[2] The sustained value of this 'confession' can be demonstrated by its iteration in recent analyses of time; see, e.g., Saro Palmeri, "Time and History, from a Singular Perspective Exploring Einstein's 'Now' Conundrum," in *KronoScope* 15, no. 2 (2015): 179–190: "Time is the most difficult and elusive concept to grasp. As with the concept of God, time is both everywhere and nowhere to be found" (p. 179).

exists in the past, present, or future (time 'will be,' 'was,' or 'is'), that is "continuous," denotes "the measure of change" (Aristotle, *Physics* IV: 10–15), and has a duration (it is 'long' or 'short'), we lack words to define it. Although time can be observed, quantified, and expressed in all human languages (cf. Klein and Li 2009), it remains largely enigmatic. Time appears to be a phenomenon whose existence we understand intuitively and that seems to exist a priori. As suggested by Immanuel Kant, it can be conceived as "the subjective condition necessary by the nature of the human mind for the coordinating of any sensible objects among themselves according to a certain law" (*Kant's Inaugural Dissertation of 1770*, trans. Eckoff, 1894: 61).

The paradox of the being and non-being of time has long been recognised (cf. Ricoeur 1984). Prompted by J.M.E. McTaggart's claim of *The Unreality of Time* (1908), Julian Barbour, for instance, argues for a timeless quantum theory of the universe and rejects the very existence of time. Instead he claims that we move between various instants of time, between various "Nows"—a conception that resonates in a surprising way with the German sociologist Niklas Luhmann's observation that every action inevitably takes place in the present. Seen from this perspective, all of humankind can be described as being tied together by a common time-horizon, a kind of endless 'Now.'[3] It is fascinating in its own right to see how the reflection about time finally leads to a convergence of approaches that unite the perspectives of social sciences and cultural studies on the one hand, and 'hard' or 'natural' science on the other: assuming that the theory of time emerged from timelessness, Barbour contends, "the task is not to study time, but to show how nature creates the impression of time" (1999: 17).

To the extent that there is no authoritative definition, 'time' might indeed be characterised as an irreducibly unknowable, impenetrable phenomenon, which continues to spark new studies across multiple disciplines that explore the perception as well as the history and philosophy of time,[4] its representation in the arts (see Clune 2013), and its transformation in the digital world (Potts 2015). Although the enduring interest in the nature of time already points to its appeal, one aspect has not yet

---

[3] Cf. from a historian's perspective Landwehr (2012: 1–34). Landwehr particularly refers to Luhmann (1993: 98–101).

[4] See Keightley (2012), Falk (2010), Callender (2002), Butterfield (1999), also Callender (2011), Gimmler et al. (1997).

been explored in detail: the fascination with 'unknown time' rather than 'time as the unknown.' By focusing on this particular aspect, the present volume does not so much aim to reinforce the widespread claim that 'time' as human construct is an essentially enigmatic phenomenon, an empty centre, which is filled or measured by devices invented by human beings (and usually communicated by the use of spatial metaphors). Instead, it explores the intense attraction that is frequently exerted by various forms of what can be referred to as 'unknown time.' The individual contributions to this volume investigate the impact of this 'unknown time' on different ways of (narrative, religious, historical, and cultural) world-making (cf. Goodman 1978), of an ordering and reordering, a composition and decomposition of familiar notions, categories, and measurements of time.

Ways of Time-Making

To clarify the approaches and perspectives presented in this volume, it is necessary to first challenge the widespread claim that time itself, as a phenomenon, is fundamentally unknown, because it seems to neglect the multiple ways in which time has been represented and conceptualised across the ages. Furthermore, this claim appears to be counterintuitive, as many people undeniably have the everyday experience that time 'exists' and is known to the extent that it can be measured ('clock time') and felt ('perceived time'). Many cultures have a strong sense of a past, present, and future: they assess the time that passes through various devices, from the hourglass to hyper-accurate atomic clocks, they experience time as 'slow' or 'fleeting,' long or short, and conceive it as sequential, quantifiable, and deterministic (*chronos*) or circular, numinous, and serendipitous (*kairos*), respectively. Instead of defining the phenomenon itself, the following articles propose to distinguish between different types of time, such as biological time, absolute and relative time, metrical and subjective time, or cyclic and linear time. In all cases, these categorisations should, however, not be understood as ontological qualifications, but rather as descriptive devices.

Furthermore, the value attached to time seems to oppose the existence of unknown time. Especially in contemporary society, time has become a leading currency—an observation that can be confirmed by the numerous time metaphors we live by in Western culture: above all the adage that 'time is money' and thus can be spent, borrowed, lost, saved, or gained (Lakoff and Johnson 2003: 7–9). In fact, time has become scarce, a rare resource, which is carefully invested in exchange for a quantifiable output. This time, which is scrupulously assessed and assigned with caution, seems to be anything but unknown. And yet, the metaphorical concept that links time with money confirms the existence of unknown time insofar as it not only suggests similarities between time and money but also underlines their differences: in contrast to money, time is inevitably immaterial. Outside the fictional realm, time banks do not exist.

How, then, can 'unknown time' be conceptualised? Following the *Oxford English Dictionary*, time denotes "a finite extent or stretch of continued existence, as the interval separating two successive events or actions, or the period during which an action, condition, or state continuous" (*OED*, 1a). It is delineated by events, actions, or conditions and (re-)constructed in our perception of these events. 'Unknown time' might, therefore, predominantly relate to 'subjective time.' In contrast to clock time, 'subjective' or 'perceived' time depends on biological conditions, regulated by an organism's circadian rhythm and affected by age, gender, cognitive abilities, and experience (Eisler 2003: 7). These factors have an impact on our individual perception of time: the speed of time experienced from a child's and adult's point of view, respectively, differs dramatically, depending on their expectations on how time is filled. Observers of the same object will experience the duration of their encounter with that object differently, based on their (cognitive and emotional) investment in that object, their familiarity with it, the context of observation, or possible distractions.

Subjective time is not only highly unstable but is also altered by memory: the experience of duration-in-passing may differ from the experience of duration-in-retrospect, resulting in a disorientation, which makes the duration of the actual encounter unknown. Although time measurement devices support the illusion that time is predictable and therefore 'known,' these devices themselves are often used as "a materialization of some universal time sense" (Greenhouse 1996: 7). Clock time, for instance, is in fact arbitrary: twice a year, at least, it is manipulated, as dictated by daylight saving time.[5] As argued by Kevin Birth, "by deferring cognitive processes to these objects, we run the risk of diminishing our ability to think about time" (2012: 31). Clocks, Birth claims, serve

---

[5] Cf. Prereau (2005), Downing (2005), and most recently Bartky (2007: 161–200).

as deceptive, cognitive tools insofar as they shape our sense of knowing time: "Clocks are tools that tell us the time, but that obscure the answer to the question of what time is" (ibid., 37). The perception and experience of time, however, is fundamentally variable. It is determined by historical, cultural, and individual factors, and thus cannot be adequately captured by static definitions. Any study of time must therefore take into account the many different domains of experience that determine our sense of time as well as the cultural practices which contribute to the 'making' of time and open up multiple (alternative) temporalities, including notions of 'unknown time.'

Although impossible to pinpoint, time can be, and frequently has been, represented in various ways: in Western culture, in keeping with Aristotle's notion that time is the measure of motion, time has predominantly been conceived as an arrow, a horizontal vector or a continuous flow, which is directional and irreversible, thereby suggesting notions of progress and evolution. The shape of the clock and the path of its hands, on the other hand, point to the notion of circular time and the image of time as a wheel, which ties in with popular beliefs in astrology and the belief in rebirth. Although the idea of a circular time tentatively opposes the existence of 'unknown time,' linear time entails the prospect of a future, "full of things that have never been" (Rilke, *Letters*, trans. Bannart and Herter Norton, 1969: 253) and a past that—like the future—only exists in the human mind and is thus rooted in the present:

> Perhaps it might be said rightly that there are three times: a time present of things past, a time present of things present, and a time present of things future. For these three do coexist somehow in the soul, for otherwise I could not see them. The time present of things past is memory; the time present of things present is direct experience; the time present of things future is expectation. (Saint Augustine, *Confessions*, trans. Pine-Coffin, 1961: 25)

According to this assumption by Augustine, both the past and the future cannot fully be grasped. Insofar as the past can only be 'made present' by means of (individual or collective) memory, whereas knowledge about the future relies on predictions, based on present memories and experiences (Bühler and Willer 2016), both are essentially unspecified or unknown. At the same time, not only the past, but especially the future, is "overdetermined" (Rosenberg and Harding 2005: 3), "conditioned by a knowledge of, and even a nostalgia for, futures that we have already lost" (ibid., 4):

> Our lives are constructed around knowledges of the future that are as full (and flawed) as our knowledges of the past. Often these future knowledges are profoundly freighted, since they involve anticipatory hopes and fears. As one commentator recently put it, our futures are junkyards of memories we have not yet had. (ibid.)

Past, present, and future thus constitute arenas that are intimately entangled and invite us to reflect upon new concepts of time, which allow us to better grasp their complex and manifold interactions. It is these processes of '(re)-composing' and '(re)-making' time, of (re-)constructing, managing, and also filling 'unknown time' as well as the appeal attached to notions of 'unknown time,' that are at the core of this volume.

These complex ways of time-making can be further examined by drawing on and slightly amending the different levels that the philosopher Nelson Goodman defined for world-making processes in a more general perspective, that is, (a) composition and decomposition, (b) weighting, (c) re-ordering, (d) deletion and supplementation, and (e) deformation (see Goodman 1978: 7–17). According to Goodman, the world we have access to has been shaped by cultural conventions. Processes of world-making thus draw on already familiar world-models, which are continuously decomposed, re-ordered, and deformed. The same can be claimed for processes of 'making' (and also 'unmaking') unknown time, insofar as 'unknown time' is conceived in opposition to time as it is known to us. Similar to the construction of a (new) world, the making of unknown time "involves some extensive weeding out and filling—actual extension of some old and supply of some new material" (ibid., 14). Goodman's final category, 'deformation,' can be applied to both "corrections and distortions" (ibid., 16) of familiar time, which might assist the construction of unknown time. As Goodman suggests, artists are particularly skilled in many if not most of these processes. Although not further examined by Goodman, narratives play an important role in conceptualising unknown time: they help us come to terms with time, "the familiar stranger" (Fraser 1987), and communicate the experience of unknown time by suggesting different ways to accelerate, decelerate, and even freeze time. Contemporary time-travelling narratives (e.g., Audrey Niffenegger, *The Time Traveler's Wife*, 2003) or narrative experiments of reversing time (e.g., Martin Amis, *Time's Arrow*, 1991) confirm the enduring fascination with familiarising and defamiliarising time by exploring the potential of 'unknown time' and rewriting 'familiar time.'[6]

As argued in this volume, the growing desire to gain more profound insights into the past, the future, and also the present has not only sparked an increasing fascination with different ways of filling potential gaps in times that were considered familiar: it has also given rise to various experiments with time at the crossroads of the familiar and the unfamiliar, the known and the unknown, thereby transgressing historical, cultural, and even scientific knowledge about 'time.' Although the wonder at the yet unexplored and sheer human curiosity already gesture towards fascination, the fascination with unknown time reaches beyond this sole interest in the gaps on the great scale of cultural history.

THE FASCINATION WITH UNKNOWN TIME

Etymologically rooted in black magic (*fascinare*—'to bewitch,' 'to becry') and historically associated with the evil eye and death-darting, predominantly female agents, such as the basilisk, the cockatrice, or the Gorgon Medusa, fascination denotes a much more intense engagement in an object, event, or a person than mere interest.[7] Both fascination and interest share some common ground, insofar as they denote deep attention, but interest includes a significantly lower degree of cognitive and emotional engagement than fascination. Interest stimulates our ability to focus on, and helps sustain our attention to, specific objects; fascination is much more aggressive in its effect:

> What fascinates us robs us of our power to give sense. It abandons our 'sensory' nature, abandons the world, draws back from the world, draws us along. It no longer reveals itself to us, and yet it affirms itself in a presence foreign to the temporal present and to presence in space. (Blanchot 1982: 32)

---

[6] Especially Modernist texts questioned and reconceptualised time by exploring its social, historical, and psychoanalytical mechanisms, e.g., H.G. Wells' *The Time Machine: An Invention* (2008; orig. 1895). Cf. Tung (2015), see also Wittenberg (2013). For a comprehensive analysis of time travelling, see Nahin (1999).

[7] Cf. Hahnemann and Weyand (2009), Connor (1998), Baumbach (2015), Seeber (2012).

Following Blanchot, fascination can be conceived as the binding of the senses to one specific object, the complete absorption in the encounter, which even suspends notions of time in the experience of ultimate presence. This inexorable attraction, which verges on secular enchantment, can be associated with objects that draw us in by their extraordinary beauty, complexity, radical otherness, or even awfulness, in both senses of the term: they leave us rapt through the wonder or admiration they provoke as well as by the terror and trepidation, and sometimes perhaps the helplessness, we feel on encountering them. As already indicated, fascination implies a temporary rupture of habitual perceptual modes, a state of exception, extraordinariness, an aesthetic friction, and a radical betwixt-and-between where our critical faculties are suspended in a moment of mental paralysis, unable to process the experience and channel the contrasting emotional responses it elicits.

It is especially in a state of liminality and ambivalence, therefore, that fascination is experienced, such as in the experience on being caught in-between familiar, yet opposing categories, such as desire and dread, which trigger conflicting and irreconcilable reactions. Fascination does thus not denote an *either-or*; instead, it refers to "*internally* divided reactions" elicited by "contrary forces simultaneously at work in fascination: seduction and shame, attraction and repulsion" (Harris 2003: 17, 15). Taking this notion one step further, fascination can be conceptualised as a temporary paralysis of the senses, which results from a failure to schematise incoming information streams and from the tricking of our evaluative faculties into a cognitive trap, a disorientation that arises from an overstimulation of sensory faculties and the generation of conflicting responses.

In contemporary culture, the term 'fascination' is predominantly used with a positive connotation, but it also (still) has a deeply unsettling aspect to it. This desire for disorientation or reorientation might explain the attraction of (im)possible worlds,[8] which are based on variations and negotiations of (unknown time) and might satiate our craving for (over-)stimulation. As the following chapters show, many of these possible worlds rely upon the enduring fascination with unknown time and the longing to fill it and make it 'known'—by astrology, prophecies,

---

[8]Cf. Ryan (1992) and (2003), as well as Umberto Eco: "can one escape from the fascination of the possible worlds …?" (1994: 99).

utopian worlds, science fiction, or cultural, historiographical, and scientific practices.

As argued by several of our authors, these attempts to familiarise the unknown do not erase, but ultimately cement, the phenomenon of 'unknown time' as the foundation for these practices to emerge. To further explore the dynamics of the production and reception of 'unknown time' across a wide range of different approaches, cultural practices, and the arts is a key objective of this book. One focus is set on the techniques used for scrutinising, preserving, and even constructing the unknown with regard to time in different cultural, historical, and social contexts. Furthermore, the individual chapters aim to rationalise the enduring desire for, as well as the dangerous lure of, the unknown—a desire that goes hand in hand with changing cultures of knowledge and methodological advancements.

The link between fascination and time is by no means exhaustively described by the reference to the alluring quality of unknown time. This seemingly banal statement is in fact quite important, because, at a first glance, fascination and time seem to have little in common at all: fascination entails some kind of petrifaction, but time is always in flux. Yet, fascination and time share common ground in that both phenomena connect in their mutual focus on the present, which serves as a vantage point in the perception of time and the experience of fascination.

As already indicated, fascination denotes a state of intense presence, in which all senses are bound in the experience of an object or person. In the petrifaction it entails, time is suspended and the present moment prolonged—in specific (albeit fictitious) instances, for example, in the encounter with particular agents of fascination, such as the basilisk, to eternity. The desire to fix time, to fully experience, live in, and be able to grasp the present moment requires the transgression of familiar time. This desire points to the wish to manipulate subjective time, to transform time into timelessness, and stretch familiar time into unfamiliar and fundamentally unknown time (e.g., via temporary petrifaction), a transformation that borders on the potential of divine power, as exemplified in an episode related in Joshua 10: 12–13: the text describes how God enables his chosen people of Israel to overcome the Amorites by making the sun and moon stand still until the end of the battle. In fact, this early example from the Old Testament perfectly expresses the profoundly contradictory and paradoxical character of the chosen motif that might be at the heart of its fascination: although God makes time stand still (as expressed in the image of the sun that no longer moves), time simultaneously has to continue to enable the Israelites to take revenge against their enemies.

In addition to these inherent paradoxical dynamics, both concepts rest on contrary forces, which elicit opposite reactions: the simultaneous prompting of deep attraction and intense anxiety, which has been associated with fascination, might also be related to our ambivalent response to time. On the one hand, time provides orientation: it regulates our daily activities and structures our lives. On the other hand, time has a menacing aspect to it: it imposes limitations while it is in itself eternally moving on, unremittingly heralding the beginning and the end of human lives as it goes on its way. Aiming to reach beyond these limitations of life and death, we feel intrigued by alternative worlds (and times), which transgress these boundaries and open up unknown times that defy previous conceptions of 'time.'

According to Steven Connor, "the desire for fascination is a desire for arrest, but of a certain enlivening kind" (1998: 12). This arrest associated with fascination might also provide a means for delineating 'unknown time,' its forms and its various functions. Insofar as the fascination with unknown time embraces these two radically opposing states—arrest and flow—the endeavour of this volume includes the key ingredients of fascination and draws attention to an enticing, potentially risky, and certainly complex field of research, one which has yet to be explored.

ROADMAP: TO BOLDLY GO WHERE NO WO/MAN HAS GONE BEFORE

Inevitably, this collection cannot cover every aspect of our topic. Nevertheless, the essays in this volume showcase some of the potentially endless richness of approaches, phenomena, and materials in this area. As such we hope that the original research published here will inspire more enquiries into the relationship between fascination and unknown time. As we show in this volume, this is a fertile field for collaboration across the disciplines. To organise the following chapters, we have chosen a structure that echoes the paradoxical nature of the three basic dimensions of time which are frequently distinguished: the resulting three sections focus on 'past futures,' 'unknown presents,' and 'future pasts,' respectively.

As suggested by the contributors, the fascination with unknown time can be traced across different ages and cultures. Therefore, it can only be adequately explored through a wide (trans-)historical and trans-disciplinary lens. The analysis of this fascination, of the desire to defamiliarise time, and the changing representation of 'unknown time' in different media, provides new insight into the ways in which time was, and still is, perceived, structured, and experienced and in which notions of 'unknown' time have shaped and continue to shape culture(s), religion(s), and the arts. Exploring the fascination with unknown time, individual chapters survey its forms, (cultural) functions, and effects; they also offer insight into techniques for alienating time. The questions that are addressed include the following:

- How can 'unknown time' be conceptualised; what are its forms, functions, and effects; how is it filled, perceived, and represented; and to what extent does it interact with concepts of (unknown) space?
- How can time be familiarised or defamiliarised, made known, or unknown?
- What role does 'unknown time' play in different cultural and historical configurations and what does this entail for our perception of the cultures in question?

The first section, "Past Futures," examines unknown time from a historical perspective and is concerned with techniques to fill and to gather knowledge about a temporal dimension that is, from a rational perspective, necessarily unknown: the future. These techniques include prophecies and further means to gather insight into knowledge of divine origin, but also rationalised practices, such as astrology, whose practitioners claim that their art provides a scientific means to acquire knowledge about future events. Although the primary focus of these essays is on time, the analysis of utopian thinking in the premodern period reveals that notions of ideal time and ideal space can sometimes converge and build an inextricable complex (as Christian Hoffarth shows in his analysis of the motif of "Jerusalem"). This section unites selected case studies that offer insights into the ways in which unknown times are used to (re-)fashion the past, the present, and the future.

Anke Holdenried examines how texts from Latin Europe considered to be prophetic could be (and actually were) reappropriated in the period between ca. 1050 and 1200. This reuse of prophecy is an indication of the fascination which this material exerted on medieval readers and provides a window onto their ways of understanding the relationship between past, present, and future, that is, onto medieval 'regimes of temporality.' Holdenried's case study examines the repurposing of two types of material (Old Testament texts about the Jewish prophets and the non-scriptural, free-floating Prophecy of the Tiburtine Sibyl). She then analyses how medieval readers applied regimes of temporality across these different genres of prophetic material, in particular by deploying onto a non-Biblical text ways of understanding time learned from scriptural exegesis. Despite operating within these limits, these reader responses demonstrate diverse (and sometimes surprising) medieval attitudes to the temporal context of the events and persons encountered in prophetic material, and hence to concepts of time.

The (dis)connection between ideal time and the unknown, which ultimately conflate in a utopian vision of time, is at the centre of Christian Hoffarth's chapter on "*Wunschzeit* Jerusalem." Based on three examples—biblical scriptures, monastic writings, and particularly Joachim de Fiore's *Liber Figurarum* (late twelfth century)—Hoffarth underscores the dynamic interaction between concepts of unknown time and unknown space in medieval 'utopias.' On this exemplary basis, he develops a typology that helps explain the seminal role of the unknown in the design and representation of ideal time (and space) with regard to futures which had already been outlined and announced in the past.

In addition to the image of time as it transpires from prophecies and 'utopian' visions, a third form of unknown time emerges in accounts that are directed to the period after the end of time: in his exploration of the history of Byzantine constructions of the future, András Kraft focuses on the underlying fabric of unknown, future time in apocalyptic texts, and analyses the cognitive schemata introduced in these narratives, which prescribe or plead for the transition into a post-apocalyptic world. As suggested in his chapter on "Living on the Edge of Time," these apocalyptic notions are based on the concepts of an essentially permeable complex of past, present, and future, which transcend familiar notions of time—ultimately introducing alternative timeframes, which result from the unsettling attraction of unknown time and the persistent desire to fill these blanks with narratives that draw on, combine, and ultimately exceed familiar structures of past, present, and future time.

While the first three chapters of this section focus on large-scale concepts of distant and 'utopian' dimensions of time, Klaus Oschema concentrates on the role of time and knowledge about the future in a discourse that connects theological debates with everyday practices and desires. In his chapter "Unknown or Uncertain? Astrologers, the Church, and the Future in the Late Middle Ages," Oschema analyses a choice of late medieval astrological texts (so-called "judicia anni" and anti-astrological treatises), to demonstrate the vivid interest that late medieval individuals and groups had in their immediate future. Although explicit reflections on the fascination with the 'unknown time' that is the future remain quite rare and can mainly be inferred from polemic anti-astrological texts, the production of astrological prognostications as such clearly attests to a widespread interest. In spite of its existence and importance, the practice of astrology also entailed fundamental logical and theological difficulties: whereas theologians first and foremost underlined the undetermined character of the future to ensure the existence of free will as a precondition of individual responsibility, more philosophical debates focused on the potential predictability of future events (which would logically entail their inevitability). From the astrologers' perspective it thus became necessary to claim their capacity to foresee future events, all the while underlining a certain degree of uncertainty of their prognostications to counter accusations of heretical beliefs. In a more general perspective, and together with the three other contributions on medieval phenomena, this chapter clearly invalidates the widespread assumption that medieval cultures in Europe did not cultivate an explicit interest in ideas about a secular future.

The final chapter of this section ("'From the Unknown to the Known and Backwards': Visualisation and Conceptualisation of Palaeontological Time") takes its cue from notions of 'unknown time' in palaeontology. Analysing two concrete representational forms used by palaeontologists to conceptualise and visualise time, Marco Tamborini derives fundamental conclusions about its epistemic features as well as the role of fascination and unknown in scientific research. As this chapter shows, the idea that geological time can be understood, expressed, and ultimately constructed by means of spatial representations, as introduced by Baron Georges Cuvier, marks a first turning point in palaeontological practice.

Tamborini traces the development of various methods used to visualise 'unknown time,' focusing on two of the most important palaeontologists of the nineteenth century, Heinrich G. Bronn and John Phillips. He demonstrates that the unknown is the basic condition for palaeontological research, not only with regard to time, but also with regard to space. Although it might be true that the representation of time through spatial metaphors is more or less inevitable, even in everyday language, this technique takes on a particular value in palaeontology, as the discipline contends with the virtually abysmal phenomenon of "deep time."

The following, second section changes the perspective from 'past futures' to the manifold manifestations of 'unknown time' in the present, thereby underlining the general impression that fascination is intimately connected with the desire for presence. Questions that are addressed in this section concern contemporary strategies of communicating 'unknown time', the defamiliarisation of time, and practices of making the known 'unknown' by rearranging temporal (and spatial) elements, by combining instead of juxtaposing 'cyclical time' and 'linear time,' by traversing time, or by applying new technologies that require a reconceptualisation of familiar notions of 'time.' Museum exhibitions, time-travel narratives, the deceleration or acceleration of time in film, experimental beats in music—these are but some examples of alternative measurements that suggest the presence of 'unknown time.'

Based on the example of the ethnographic museum, Katja Wehde discusses two levels of 'unknown time.' She argues that, on the one hand, some aspects of ethnographic time are factually unknown as nineteenth-century ethnographers did not document experiences of transition and change, but sought to record what they regarded as the essence of the cultures they observed (thereby deliberately constructing their objects as being timeless and without a history in their own right). On the other hand, although some aspects actually are known that might serve to historicise the objects and their context of origin by situating them in time, museums often chose not to include this information in their displays. Such information that is usually not made available to the public frequently includes details about the collectors, about the procedures of acquiring objects, subjective experiences of the ethnographers in the field, as well as recent scholarly work on the regions represented in the museums. Starting from this distinction between retrievable and irretrievable time, this chapter investigates which time is being reintroduced to museum displays and what narrative strategies are applied to negotiate between the time that can and the time that cannot be known.

The second chapter in this section examines the nexus between 'cyclical' and 'linear time': based on ethnographic fieldwork in the public transport sector of urban Lagos—and countering Henri Lefebvre's claim that the structure of everyday life was associated with the nonaccumulative routing of 'cyclical' or 'immanent time,' all the while being averse to the permanently progressive 'linear' or 'transcendent time'—Daniel E. Agbiboa argues that 'cyclical' and 'linear time' are in fact closely intertwined. Investigating practices of repetition and the use of repetitive slogans and symbols that are displayed on public and commercial vehicles, Agbiboa claims that the interconnection of these seemingly conflicting notions of time affects and ultimately shapes the foundations of the inhabitants of Lagos, who are daily confronted with an existence that is characterised by features which are prone to evoke the impression of meaninglessness, insecurity, and a sense of not belonging to the universe. At the interfaces of conflicting concepts of time and in the interchange between destinations, a sense of unknown time can emerge, which in turn reinforces the need for a more experimental re-positioning and re-orientation in everyday life, and that finds its expression in a broad variety of slogans which rely heavily on the sphere of the divine.

The role of temporal friction and transgression of familiar concepts of time is further explored in the following chapter with regard to music and the underlying temporal structure of 'Gypsy' folklore, a type of music that combines the familiar with the unfamiliar, or exotic, and adds a further dimension to the analysis of unknown time across different media. As Anna Piotrowska argues, the stereotype of the 'Gypsy' raises very specific expectations with regard to the 'sense' of timing, that is, to both the ability to make sense or establish a (familiar) pattern and to the affective response to musical time in the experience of unknown time in musical compositions. The unknown in 'Gypsy airs,' 'Gypsy waltzes,' or 'Gypsy polkas' arises from frictions in rhythm, pauses that integrate blank spaces of the unknown, or the perpetual wandering between and across unknown spaces, which these musical pieces tend to imitate through specific sound patterns. Being constantly on the move, 'Gypsy music' is located in a limbo state, neither 'here' nor 'there,' opening up a fourth dimension of time in which past, present, and future are combined in a sense of presentness that is perpetually delayed and deferred. According to Piotrowska, the motif of 'Gypsy' folklore points to the eternal aporia of time: its perception may elude our senses, though it can never be totally denied in reference to experiencing music.

The final chapter of this section bridges the present and the future: historian Caroline Rothauge scrutinises notions of 'unknown time' in the German Reich around 1900, an era that has been regarded a caesura because of the multiple twists and turns brought about by technological innovations, such as the telegraph and the telephone, that made 'present' events at once readily accessible over large distances, but in a paradoxical way also rendered the 'present' increasingly mysterious and unknown. Rather than backing the master narrative of an all-encompassing acceleration, Rothauge argues that the unknown experienced by the contemporaries was rooted in a yet again increased pluralisation of both the notions and uses of present time(s). This discovery provoked a series of responses that combined repudiation with fascination, a combination which matches the inherent ambivalence associated with the era of high modernity.

Pursuing the complex and intertwined order of chronology, our volume concludes with a section dedicated to the analysis of future pasts that explores the unknown with regard to cinematographic time, gaming time, and literary f(r)ictions of unknown futures. Based on selected case studies, the first two chapters by Sheldon Lu and Dorothee Hou and Hauke Lehmann present in-depth analyses of representations of unknown time and the experience and emotional effect of unknown time in film. The first chapter ("The Time-Image and the Unknown in Wong Kar-wai's Film Art") sets out to explore the ways in which unknown time is introduced through the distortion of familiar time imagery and investigates techniques of accelerating and decelerating time. The second chapter ("Suspense in the Cinema") then focuses on the experience of time and the interplay between time, knowledge, and affect. Based on the assumption that suspense, as an "anamorphosis of cinematographic time" (Pascal Bonitzer), combines two temporal perspectives that seem to be mutually exclusive (the 'now' and the 'not-yet'), Hauke Lehmann explores the tension between the known and the unknown that emerges from manipulations of familiar concepts of time and the process of 'giving the viewer time' to respond to and experience the thrill of the unknown.

The final chapter in this section is dedicated to narrative constructions of future times in fiction: analysing structures of unknown and unfamiliar time in science fiction, Kai Wiegandt explores the fascination we feel when reading fictional accounts on un/known futures—accounts, which, as Wiegandt argues, provide vicarious experiences of unknown ways of being-in-the-world and thus relate to the desire of presence, which is projected into the future. As Wiegandt demonstrates, concepts of 'futurity' greatly contribute to the fascination with science fiction. This chapter not only provides a concise historical overview of the developments of the genre with regard to the unknown, making a distinction between hard and soft science fiction; it also offers in-depth analyses of the ways in which science fiction reflects upon present (and thus familiar) scenarios in estranged forms to prompt readers to decode analogies between future narratives and present experiences and engage in alternative forms of world-making that draw heavily on timeframes that capitalise on the fascination with the unknown.

BIBLIOGRAPHY

Aristotle. 1996. *Physics*, trans. Robin Waterfield. Oxford: Oxford University Press.

Barbour, Julian. 1999. *The End of Time: The Next Revolution of Physics.* Oxford: Oxford University Press.

Bartky, Ian R. 2007. *One Time Fits All: The Campaigns for Global Uniformity.* Stanford: Stanford University Press.

Baumbach, Sibylle. 2015. *Literature and Fascination.* Basingstoke: Palgrave Macmillan.

Birth, Kevin K. 2012. *Objects of Time: How Things Shape Temporality.* New York: Palgrave.

Blanchot, Maurice. 1982. *The Space of Literature*, trans. Ann Smock. Lincoln: University of Nebraska Press.

Bühler, Benjamin, and Stefan Willer (eds.). 2016. *Futurologien: Ordnungen des Zukunftswissens.* Paderborn: Wilhelm Fink.

Butterfield, Jeremy (ed.). 1999. *The Arguments of Time.* Oxford: Oxford University Press.

Callender, Craig (ed.). 2002. *Time, Reality and Experience.* Cambridge: Cambridge University Press.

——— (ed.). 2011. *The Oxford Handbook of Philosophy of Time.* Oxford: Oxford University Press.

Clune, Michael. 2013. *Writing Against Time.* Stanford: Stanford University Press.

Connor, Steven. 1998. Fascination Skin and the Screen. *Critical Quarterly* 40 (1): 9–24.

Downing, Michael. 2005. *Spring Forward: The Annual Madness of Daylight Saving*. Washington, DC: Shoemaker & Hoard.

Eco, Umberto. 1994. *The Limits of Interpretation*. Bloomington: Indiana University Press.

Eisler, Anna D. 2003. The Human Sense of Time: Biological, Cognitive and Cultural Considerations. In *The Nature of Time: Geometry, Physics and Perception*, ed. Rosolino Buccheri, Metod Saniga, and William Mark Stuckey, 5–18. Dordrecht: Kluwer Academic.

Ende, Michael. 1973. *Momo, oder: die seltsame Geschichte von den Zeit-Dieben und von dem Kind, das den Menschen die gestohlene Zeit zurückbrachte. Ein Märchen-Roman*. Stuttgart: Thienemanns Verlag.

Falk, Dan. 2010. *In Search of Time: The History, Physics, and Philosophy of Time*. Basingstoke: Palgrave Macmillan.

Fraser, Julius Thomas. 1987. *Time, the Familiar Stranger*. Amherst: University of Massachusetts Press.

Gimmler, Antje, Mike Sandbothe, and Walter Ch. Zimmerli (eds.). 1997. *Die Wiederentdeckung der Zeit: Reflexionen, Analysen, Konzepte*. Darmstadt: Primus-Verl.

Goodhew, Linda, and David Loy. 2002. Momo, Dogen, and the Commodification of Time. *KronoScope* 2 (1): 97–107.

Goodman, Nelson. 1978. *Ways of Worldmaking*. Indiana: Hackett.

Greenhouse, Carol. 1996. *A Moment's Notice*. Ithaca: Cornell University Press.

Hahnemann, Andy, and Björn Weyand (eds.). 2009. *Faszination: Historische Konjunkturen und heuristische Tragweite eines Begriffs*. Frankfurt: Peter Lang.

Harris, Oliver. 2003. *William Burroughs and the Secret of Fascination*. Carbondale and Edwardsville: Southern Illinois University Press.

Kant, Immanuel. 1894. *Kant's Inaugural Dissertation of 1770*, trans. William J. Eckoff. PhD Dissertation, Columbia College, New York.

Keightley, Emily. 2012. *Time, Media, and Modernity*. Basingstoke: Palgrave Macmillan.

Klein, Wolfgang, and Ping Li. 2009. *The Expression of Time*. Berlin: De Gruyter.

Lakoff, George, and Mark Johnson. 2003 [1980]. *Metaphors We Live By*. Chicago: University of Chicago Press.

Landwehr, Achim. 2012. Von der 'Gleichzeitigkeit des Ungleichzeitigen'. *Historische Zeitschrift* 295: 1–34.

Luhmann, Niklas. 1993. *Soziologische Aufklärung*, vol. 5. *Konstruktivistische Perspektiven*, 2nd ed., 98–101. Opladen: Westdeutscher Verlag.

Nahin, Paul J. 1999. *Time Machines: Time Travel in Physics, Metaphysics, and Science Fiction*, 2nd ed. New York: Springer.

Nünning, Vera, Ansgar Nünning, and Birgit Neumann (eds.). 2010. *Cultural Ways of Worldmaking*. Berlin: De Gruyter.

Palmeri, Saro. 2015. Time and History, from a Singular Perspective Exploring Einstein's 'Now' Conundrum. *KronoScope* 15 (2): 179–190.

Prereau, David. 2005. *Seize the Daylight: The Curious and Contentious Story of Daylight Saving Time*. New York: Thunder's Mouth Press.

Potts, John. 2015. *The New Time and Space*. Basingstoke: Palgrave Macmillan.

Ricoeur, Paul. 1984. *Time and Narrative*, vol. 1, trans. Kathleen McLaughlin and David Pellauer. Chicago: The University of Chicago Press.

Rilke, Rainer Maria. 1969. *Letters of Rainer Maria Rilke: 1892–1910*, trans. Jane Bannart Greene and M.D. Herter. New York: Norton.

Rosenberg, Daniel, and Susan Harding. 2005. Introduction: Histories of the Future. In *Histories of the Future*, ed. Daniel Rosenberg and Susan Harding, 3–18. Durham: Duke University Press.

Ryan, Marie-Laure. 1992. *Possible Worlds, Artificial Intelligence, and Narrative Theory*. Bloomington: Indiana University Press.

Ryan, Marie-Laure. 2003. *Narrative as Virtual Reality: Immersion and Interactivity in Literature and Electronic Media*. Baltimore and London: Johns Hopkins University Press.

Saint Augustine. 1961. *The Confessions*, trans. R.S. Pine-Coffin. New York: Penguin.

Seeber, Hans Ulrich. 2012. *Literarische Faszination in England um 1900*. Anglistische Forschungen, vol. 426. Heidelberg: Winter.

Tung, Charles M. 2015. Modernism, Time Machines, and the Defamiliarization of Time. *Configurations* 23 (1): 93–121.

Wells, H.G. 2008 [1895]. *The Time Machine: An Invention*, ed. Stephen Arata. New York: Norton.

Wittenberg, David. 2013. *Time Travel: The Popular Philosophy of Narrative*. New York: Fordham University Press.

PART I

Past Futures

CHAPTER 2

The Old Made New: Medieval Repurposing of Prophecies

Anke Holdenried

In the Middle Ages there was intense interest in predictions about the future. This is not merely because there is a natural human desire to know what is going to happen next, so one can prepare for it. In the medieval period, ideas about the future were framed within a specifically Christian context that believed history would culminate in Judgement Day. One of the peculiarities of medieval thinking about the future is therefore that it was conceived as a time which was simultaneously 'unknown', because it was yet to come, and 'known', because its endpoint at the Second Coming of Christ was foretold in the Bible.

Here I focus on the fact that many of the predictions popular in the Middle Ages were not novel creations but in fact had previously appeared in a different form in older texts but were then recycled into a new context. This reuse of 'past' prophecy necessarily involved the reinterpretation of existing material. In short, it required the repurposing of this material. This repurposing can be observed across many different genres, from free-floating texts predicting the future to the books

A. Holdenried (✉)
Department of Historical Studies, University of Bristol, Bristol, UK

© The Author(s) 2017
S. Baumbach et al. (eds.), *The Fascination with Unknown Time*,
DOI 10.1007/978-3-319-66438-5_2

of the Bible. It applied regardless of whether such texts were written predominantly in predictive mode or with a more historical orientation.

In this essay I examine the ways in which prophecy was repurposed, chiefly in the period ca. 1050–1200 on the basis of a case study. I argue that the reinterpretation of older prophetic texts was not just about keeping them current. Instead, my paper treats their reuse as a window onto medieval approaches to time. The reception of both biblical and non-biblical prophetic texts required a particular type of reinterpretation: readers had to move narratives from one temporal context into another. In each case, as they repurposed their materials, readers had to find ways to connect anew with the future set out therein. How did they do this? This question is at the heart of the present discussion.

We must begin by defining some terms and concepts. Every period and culture has its own way of understanding the relationship between the past, the present, and the future, that is, between historical time, current time, and times yet to occur. In modern historiography, the relationship among these three categories (or levels) of time is often called a 'temporality'. The way a society approaches this relationship is then referred to as a 'regime of temporality'. Although this latter phrase has been much deployed recently in studies of time, unhelpfully, there does not seem to be a single agreed definition of it.[1] Here I follow Jordheim, who has reviewed current usages of the term and conceives it as a "set way of understanding and dealing with time according to which the relationship between the past, present, and future, and thus the direction, speed, and rhythm of history, can be defined" (2014: 501).[2] That is, regimes of temporality are ways of understanding time.

While I find Jordheim's definition helpful, it is an umbrella term. Other scholars have offered their own particular examples of regimes of temporality tailored to the material they discuss. Unfortunately none of the regimes of temporality in the existing historiography fully captures the issues considered in this paper. For present purposes therefore

---

[1] The term 'regime of temporality' (or, often, synonymously the term 'temporal regime') developed around Francois Hartog's notion of the 'regime of historicity'. Hartog uses that phrase in his reflections on how history is experienced, conceived of, and written down in different periods, see Hartog (2015: 8–9 and 106). For further discussion of the term 'regime of temporality', see Jordheim (2014: 498–518); see also n. 3, below.

[2] See also Jordheim (2014: 499–501 and 509).

I propose two other regimes to which I shall refer as the *chronological* regime of temporality and the *synchronous* regime of temporality.

In the chronological regime, events are understood to have occurred in the past, are now completed, and have no further ongoing significance. By contrast, the synchronous regime considers that past events may still be significant as symbols, metaphors, allegories, or the prefiguration of later events. Under the synchronous regime, therefore, the significance of past events survives the passage of time; this contrasts with the chronological regime of temporality. Both regimes (and the distinction between them) are rooted in their approach to the reading and interpretation of texts. I say more about this later. I have adopted these terms—the synchronous and the chronological—to emphasise the attitude of medieval readers to the temporal context of the events, persons, gestures, etc. they consider.

As a way of discussing these regimes of temporality, in the following I explore the repurposing of two distinct types of material not normally considered together: first, Old Testament narratives about the Jewish prophets, and second, the anonymous Prophecy of the Tiburtine Sibyl, a prediction with roots in late antiquity. In this essay, I juxtapose their reception in a common intellectual environment. The nature of that environment will emerge below.[3]

My approach offers a new way of examining free-floating predictive narratives: my primary concern is not so much with how predictive texts function as a vehicle to express social, political, and religious crises as with the broader conceptual toolkit with which medieval people approached free-floating prophecy. In particular, I want to explore the role illustrated by the "set ways of understanding time" (that is, the 'regimes of temporality') that had formed in readers' minds as a result of reading and interpreting scripture in other contexts.

Erich Auerbach was among the first scholars to emphasise that the techniques of reading and interpreting scripture were so culturally pervasive that they "often enter[ed] into the medieval view of everyday

---

[3] This approach also departs from current studies of medieval reworkings of the Tiburtina (and its extensive manuscript transmission), which, although they acknowledge the impact of the liturgy, concentrate not on regimes of temporality but on the role of memory in shaping medieval approaches to the text, see Holdenried (2006: 111–126).

reality" (1984: 61).[4] He did much to clarify the cultural influence of exegesis in his seminal paper "Figura" (ibid.).[5] Yet, despite this work's immense contribution to our understanding of Judeo-Christian temporal concepts and structures, regrettably, it is often absent from the bibliographies of subsequent studies devoted to 'Time'.[6] It has much to tell us, however, about foundational epistemological experiences with prophecy in the medieval world, because Auerbach stressed in particular the technique of reading the Old Testament *figuratively*, that is, as a foretelling of real people and events in the Gospel. For example, he stated that "the naming of Joshua-Jesus is a *Realprophetie* or 'phenomenal prophecy' or prefiguration of the future saviour" (Auerbach 1984: 29).[7] In using the term *Realprophetie* Auerbach talks about the figural realism of the Old Testament whereby a real (i.e. historical) event, person, object, or gesture in it prefigures another, later one that is also a historical reality. This approach rested on the belief that all history was merely a component part of the greater history of salvation.

This belief made it difficult for the Bible's interpreters to distinguish its phases by applying the categories of 'past', 'present', and 'future' to scripture. As Auerbach noted, Bible hermeneutics reflects this: according to its rules, events and people in the Bible could be read either literally (as historical occurrences), or allegorically. The allegorical approach treats the whole Bible as a kind of hyper-extended metaphor (or collection of metaphors) whose placement along the chronological continuum

---

[4] Originally published in German in 1938, also reprinted in Auerbach, *Gesammelte Aufsätze zur romani schen Philologie* (1967): 55–92. In this essay I cite the English translation by Ralph Mannheim (see Auerbach 1984: 61). For "Figura's" continued importance, see Balke and Engelmeier (2016) and Auerbach (2014), with an introduction by James I. Porter.

[5] For an appreciation of Auerbach's impact, see Lerer (1996).

[6] See, for example, Hunt (2008), Munn (1992) (with "Notes on the Future" on pages 112–116), and Burke (2004). Auerbach's work is absent, too, from Koselleck's seminal study of the history of time in which the analysis of language alongside that of philological and hermeneutical pardigms and practices plays a key part, see Koselleck (2004), first published in German as Koselleck, *Vergangene Zukunft: Zur Semantik Geschichtli cher Zeiten* (1979).

[7] Auerbach's translator, Ralph Manheim, translated the German word *Realprophetie* as 'phenomenal prophecy', a somewhat curious choice. I assume that it is not meant in the sense of 'remarkable, outstanding', but is related to the term 'phenomenon', that is, meaning 'fact' or 'occurence' (and perhaps to the philosophical term 'phenomenological'), see Auerbach (1984: 29–34).

was not important per se because everything referred either back or forward to something else. In terms purely of exegetical technique, the allegorical approach in turn could be subdivided further, so Old Testament events could be viewed typologically (as prefigurations of other events and people, typically those to appear in the New Testament) or by allegorising them further (to give them some other spiritual sense). A scriptural passage, therefore, might be given either a literal reading, as a straightforward narrative of an event that had occurred in the past but that had no further significance (placing it in my terms in the chronological regime of temporality), or a typological/allegorical reading, giving the episode some continuing contemporary or future meaning (in my terms, putting it in the synchronous regime of temporality). Note that, in medieval terminology, this allegorical or metaphorical approach was called a 'spiritual understanding' of scripture.[8] For the purposes of this paper, allegorical, typological, and figural exegetical approaches are treated the same. There are differences between these approaches and in other circumstances they may have different meanings and be applied in different ways, but not in the context of my argument here.[9]

For now, I merely note that the Old Testament pre-figurings, which Auerbach called "*figurae*" or "types," found later fulfilment and hence meaning in the Gospel. Nevertheless, this new meaning was only provisional: as Auerbach reminds the reader ultimately 'figures' are "tentative forms of something eternal and timeless" (1984: 59). In this sense, they demand a metaphorical, allegorical, or figural/typological rather than a literal interpretation. Consequently, these modes of interpretation embody a flexible attitude to the temporal concepts 'past', 'present', and 'future', since for the purposes of textual interpretation all can be understood as being simultaneously significant (in my terms, putting them in the synchronous regime of temporality).

---

[8] See, for example, "per mysterium spiritalis intellectus" and "per spiritalem intellegentiam", n. 15, below.

[9] In this essay I use either the conjoined form *typological/allegorical* or *figural/typological* because "in figural interpretation one thing stands for another, since one thing represents and signifies the other, [so] figural interpretation is 'allegorical' in the widest sense" (Auerbach 1984: 53–54). However, Auerbach also observes that sometimes typology and allegory can be treated as different, for while typology is anchored in concrete events, allegories are often ethical or mythical interpretations which may not be historical and are not historically anchored in concrete events (whether past, present, or future).

This attitude first arose within the narrow confines of Bible study but soon became a widely applicable habit of mind. It also, as Auerbach noted, "provide[d] the medieval interpretation of history with its general foundation" (ibid., 60–61). Other scholars have considered this relationship between figural/typological interpretation and forms of history writing in the Middle Ages.[10] Here I shall consider the pervasive cultural influence of exegesis in relation to a different genre, that of non-scriptural prophecy. I shall do this by asking how patterns in reader responses to one example of this genre (the Sibylla Tiburtina) map onto the regimes of temporality connected to biblical exegesis. My particular interest lies here with the chronological and synchronous regimes associated with the interpretation of Auerbach's *Realprophetie* (that is, with the Old Testament: it provided the vast majority of *Realprophetie* in the Bible).[11]

In order to approach the reception and repurposing of biblical texts I begin with a letter of spiritual advice written by Peter Damian in 1069.[12] Since later I shall turn to the Sibyl, note that this letter is almost contemporary with the earliest surviving manuscript of the Prophecy of the Tiburtine Sibyl, dated 1047.[13] In the letter Peter Damian presents an allegorical treatment of the Old Testament Exodus narrative, that is, of the journey of the Hebrew people out of Egypt led by Moses and Aaron. Peter's focus is on the people's 42 rest-stops in the desert, which he allegorises as so many stages in the interior transformation of a monk striving towards spiritual perfection.[14]

Peter championed the view that Exodus had an ongoing significance, giving us an excellent example of the repurposing of scripture based on the belief that it represented a *Realprophetie*, that is, he took an allegorical approach. He looked upon the Exodus narrative as

---

[10] See, for example, Boynton (2006) and Spiegel (2016). See also (in this volume) Hoffarth and Kraft. On a different albeit related note Wiegandt (in this volume) illustrates allegory's continuing importance in the modern period for articulating the relationship between different categories of time.

[11] Other important strands in the study of time in the medieval period cannot be considered here, for example, 'social time': Adam (2004), Burke (2004), 'measuring time': Borst (1993), Stern (2003), 'time and creation': Dales (1990).

[12] Peter Damian (2005: 103–130 [Letter 160]) (translated into English by Owen J. Blum and Irven M. Resnick). For the Latin text, see Reindel (1993: 100–134).

[13] See Holdenried (2016).

[14] See Ex 12:35–17:2; the 42 rest-stops have their source in Nm 33:3–48. Peter's exegesis of them relies on Jerome, "Letter 78" (see Peter Damian 2005: 110, n. 32).

totally fulfilled for us through the mystery that underlies our spiritual understanding. For whatever then occurred visibly, is adapted to our need by *allegorical interpretation*, as the age long past is made to serve us at the present time. ... We too came forth from the ordeal of Egyptian servitude, and strive to enter the promised land by many stages, that is, by varied advancement in virtue. (Peter Damian 2005: 107 [9] [my emphasis])[15]

Peter's statement that Old Testament events are 'made to serve us at the present time' implies that even as time passed events remained significant (at least at the metaphorical level); they remained 'in play', that is, they could have a *current* meaning regardless of when they occurred in the past. Events such as the Flood, Exodus, the Babylonian Exile, etc. were all equally current:[16] that is, although separated from each other and from the moment Peter wrote by periods of time of varying length, from his perspective all were now simultaneously significant regardless of when they occurred in the past. For this reason, I have dubbed this approach the synchronous regime of temporality: all these events, people, objects, and so forth had (and continued to have) simultaneous significance.

There was good authority for Peter's view: the Apostle Paul said the events of the Exodus befell the Jews "as figures" so that Christians should not lust after evil things (Cor 10:6) (Vulgate).[17] But Peter Damian was also well aware that not everyone shared his view of the story of the Jews in the desert. Elsewhere in his letter, he mentions "querulous person[s],"

---

[15] Translated by Blum, who renders the two very similar phrases *spiritalis intellectus* and *spiritalem intellegentiam* by two very different English phrases ('spiritual understanding' and 'allegorical interpretation'). On this point, I agree with the translator because application of *spiritalis intelligentia* involves allegory; see van Liere (2014: 114–115). Cf. Reindel (1993: 104–105 [my emphasis]): "[Notandum autem quoniam omnis ille discursus et quicquid illic gestum hystorialiter legitur,] totum in nobis *per mysterium spiritalis intellectus* impletur. Quod enim tunc visibiliter gestum est, nobis *per spiritalem intellegentiam* congruit, nostro tempori vetus illud saeculum militavit. Haec enim, ut ait apostolus, 'in figura contingebant illis'. Nos enim de fornace Aegyptiacae servitutis egredimur, et terram repromissionis ingredi per plurima mansionum loca, hoc est per per diversa virtutum incrementa conamur".

[16] Cf. András in this volume, pp. 71–74.

[17] ... "haec autem in figura facta sunt nostri ut non simus concupiscentes malorum sicut et illi concupierunt." This passage had also been noted in Peter's source: Jerome's "Letter 78," see n. 18, below.

as well as "some people who are ignorant of God's plan." Peter does not identify them (*Letters* 2005: 107 [8]).[18] Whoever they were, however, regarding their views, Peter expresses the concern that

> [the] people who are ignorant of God's plan argue that it is frivolous and superfluous to read the account of these rest stops in the church [i.e. the Exodus account of the wandering in the desert]. For they are of the opinion that knowing or reading about this matter serves no useful purpose whatsoever, thinking that [Old Testament] history narrated only what has happened, and that this event has now passed away with age, and that today it should have no further interest for us. (Peter Damian 2005: 106 [7])[19]

Peter Damian here reports an instance in which the past was perceived in terms of separateness and discontinuity. In Peter's letter this particular perception of the past as a time cut off from the present implies the use of the technique of literal exegesis whose purpose is to recover the historical meaning of the text (the "Old Testament ... narrates what has happened"[20]). We cannot say with certainty whether the perception came first (so that the choice of exegetical technique reflects it) or vice versa (that is, that the application of literal exegesis resulted in a sense of discontinuity and separateness from the past). Whatever the case, together this perception of separateness from the past and the literal exegetical technique fit into what I referred to earlier as the 'chronological' regime of temporality.

---

[18] Cf. Reindel (1993): "Agrediar ergo, frater mi, si tibi onerosum non est, mansionum illarum figuras summatim ac succincte perstringere, et quod ex dictis patrum indagare potuerim, compendiosis verbis breviter annotare, ut *querelosus* quispiam ex gustu micarum labentium colliget, quam nectareis dapibus pleni ferculi mensa redundat" (p. 104 [my emphasis]). On those who are ignorant of God's plan, see n. 19, below. Note that Peter's source (Jerome, "Letter 78") makes no mention of such querulous persons, nor of their approach to the Old Testament, so this must be Peter Damian's observation about his own time, see Jerome, *Sancti Eusebii Hieronymi Epistulae*, ed. Hilberg (1970: ep. 78).

[19] Cf. Reindel (1993: 103): "... cum nonnulli divinae rationis ignari frivolum conquerantur atque superfluum, ut in aecclesia legatur istarum descriptio mansionum. Arbitrantur enim haec scire vel legere nil penitus utilitatis afferre, putantes quod rem tantummodo gestam narret hystoria, et hanc cum ipsa tunc vetustate transisse, neque nunc ad nostram aliquatinus notitiam pertinere".

[20] See n. 19, above.

Ironically, to clarify my understanding of the chronological regime of temporality I can only offer a metaphorical explanation of the term, although in my defence, I note that Lynn Hunt has commented:

> Time ... requires metaphor. It flows like a river, accelerates like an engine, flies like a winged chariot, freezes like instant ice, stands still like a heart between beats. ... Without the metaphors ... the fourth dimension would be exceedingly difficult to grasp. Linguists have noted that it is virtually impossible to talk about time without invoking motion (wiggling skirts, engines, chariots, arrows) and spatial content (short, long). (2008: 1)

So perhaps the addition of another metaphor to the stock may be of value. When I refer to the 'chronological' regime of temporality I have in mind a particular linear conception of time, defined by its forward-looking direction of gaze. In other words, this regime requires that time is imagined as a river, or, as I prefer to think of it, similar to the mechanism of a ratchet (a tool that can be turned in only one direction but not the other).[21]

By analogy, in the 'chronological' regime of temporality, the 'present' is seen as one click in a continuous series of clicks on time's ratchet, which is capable of moving only forward towards the future, with no possibility of a backward motion. This 'ratchet' view characterises the chronological regime of temporality: it implies that both 'present' and 'future' time are sharply separated from the 'past'. This view of time is that taken by the querulous persons to whom Peter Damian refers; he castigates their position as "insane" and "nonsense."[22]

This criticism of the chronological regime of temporality, and hence of the literal exegetical technique, necessarily implies a passionate endorsement of its opposite, the typological/allegorical mode of interpretation and hence of the synchronous regime of temporality. This is no surprise because modern scholarship generally considers this mode

---

[21] Both images (river and ratchet) imply forward motion, but the image of the flowing waters of a river makes it harder to isolate and locate specific moments, i.e., to pinpoint the 'present'.

[22] Cf. Reindel (1993: 103–104): "Sed si suptiliter ipsa scripturae verba perpendimus, quam extremae dementiae sit hoc dicere, luce clarius invenimus. ... Et quis hoc audeat dicere, immo quis temerario praesumat ore garrire, ut quod Domino iubente conscribitur, nil utilitatis, nulla conferat emolumenta salutis?".

to have been culturally pervasive in the Middle Ages; indeed, the syn-chronous regime of temporality that flows from the allegorical mode (and which blurs the distinction between the temporal categories of 'past', 'present', and 'future' because the significance of past events sur-vives) is seen as the default setting of medieval culture. Peter Damian's observations, however, remind us that although he preferred the syn-chronous regime of temporality there was another one, which he rejected, the chronological regime of temporality. As an aside, note that Peter's letter also shows that medieval individuals clearly possessed the mental tools to think in a manner that contemporary scholars have labelled as 'modern' but which medieval scholars simply called "igno-rant."[23] This note challenges the way 'modern' temporalities are con-ceived and labelled. However, further discussion of this lies outside the scope of this essay. Returning to the issue at hand, recall that although he disapproved of it, Peter did at least recognise the existence of alter-native ways of interpreting the Exodus narrative with different temporal perspectives on past events.

I now turn from the repurposing of a scriptural text considered to be prophetic (Exodus) to approaches to a non-scriptural predictive text. Often the very *raison d'être* of such texts (that is, to illuminate the future) depended on how they used the temporal spectrum of past, pre-sent, and future. This concept is certainly true in the case of a Christian eschatological narrative we know today as the Sibylla Tiburtina (hence-forth Tiburtina). Before I attend to the temporal structures embedded in this work, some summary comments on its dissemination and evolution are required. Anyone seeking to understand this has to unravel roughly five centuries of textual history to unpick its fourth-century core from the medieval text preserved in the earliest eleventh-century manuscript, a time span providing ample opportunity for the deposition of differ-ent layers of amendments.[24] For example, at least four Latin versions

---

[23] See Peter Damian, cited above, n. 19. The sharp separation of the present from the past which characterises the chronological regime of temporality is often regarded as the defining feature that marks a distinctively 'modern' understanding of time; see, for exam-ple, Koselleck (2004). For further discussion of the label 'modern' in relationship to tem-porality, see Spiegel (2016) and Jordheim (2014): especially p. 506.

[24] See Holdenried (2006: 131–146) and, for a specialist discussion of the Tiburtina's pre-manuscript history (c. 400–1000), see id. 2014. See Holdenried (2006: 231) for a full bibliography of printed editions of the Sibylla Tiburtina.

survive, showing that the reuse of prophecy was not merely—in the worst sense—mechanistic recopying. Rather, as the text migrated from the Eastern part of the Empire to the medieval West, 'the future' which it describes was reinterpreted to apply to other people and other situations. In the Tiburtina's literary history between ca. 400 and 1050 we can, therefore, observe a constant repurposing of the prophecy by anonymous redactors.[25] This observation in turn requires the application of regimes of temporality, that is, of the set ways of understanding and contending with time that served to define the direction of history and the relationship between the past, the present, and future.

To better understand the relationship between medieval regimes of temporalities and the Tiburtina, let us first consider the text's temporal structure. As one of its framing devices, the narrative is structured as a world chronicle presenting a narration of historical events spread over the course of nine ages, or, in the text's own terminology, *generationes*.[26] This historical narration starts with events in ancient Rome, includes an account of the birth and passion of Christ, then narrates (more or less chronologically) the deeds of various kings and emperors, and finally ends with a description of Judgement Day at the End of the World when Christ returns. Summarily, the historical narration included in the Tiburtina mentions secular and sacred events in ancient Rome, the rulers of Egypt, the *rex grecorum*, and Lombard, Carolingian, Ottonian, and Hohenstaufen rulers; and also, still within historical time (albeit a historical time yet to come), a last ruler who will defend mankind against the Antichrist before abdicating at Christ's second coming.

In the Tiburtina, then, time is understood as a series of real, concrete events occurring across the temporal trajectory of past, present, and future. The text also describes a universal future for all mankind, culminating in Christ's return and the End of the World. On a meta-level this narrative thus expresses the dominant idea of the medieval period, that the flow of time was identical with the history of salvation which would come to an end with Judgement Day. The text's location within this meta-timeframe is reinforced by the associations of the very striking and memorable acrostic poem that concludes the Tiburtina's narrative.

---

[25] This reframing took various forms, some of which we can only reconstruct hypothetically because the surviving manuscript evidence is all post-1047, see Holdenried (2014).

[26] The Tiburtina's terminology here also mimics genealogical ways of ordering time, see, for example, Gallois (2007: 110–121).

The acrostic depicts Judgement Day, a subject matter which puts the poem into the meta-timeframe because it describes the culmination of salvation history. However, this was not the poem's only context, because these acrostic verses also had a life separate from the rest of the Tiburtina as part of the readings from the Christmas liturgy. I return to this below.

For now, let us note that the temporal interpretation of the Tiburtina was not static. As the Tiburtina circulated it had to be constantly interpreted afresh because the events it 'predicted' consistently failed to occur. As mentioned, this reinterpretation was expressed by amendments to the text itself. These revisions give an insight into this process of reinterpretation. For current purposes, this is best illustrated by the Tiburtina's regnal list, a sequence of rulers that culminated in the Last Emperor, destined to fight the Antichrist. Earlier scholars often assumed that the text's king list was constantly updated so as to suggest that the Tiburtina's prophecies were always on the verge of fulfilment. However, to judge by the text's surviving textual variations (including headings and marginalia), many of its scribes and readers were ambivalent whether to regard the prophecy as being about the past, about the present, or about the future. In particular, reactions to the text were very seldom shaped by the expectation that a current ruler would fulfil the role of a last Emperor as predicted in the Tiburtina. I have found little evidence that anticipation of this particular part of the events of the future generated changes to the narrative. However, the king list virtually ceased to be updated in manuscripts after the twelfth century. From that point on it seems that the text was no longer read as if the text's expectation of a last ruler would be realised imminently in the 'now.'[27]

Indeed, somewhat surprisingly, given that we are working with a seemingly predictive narrative, my study of the Tiburtina and its medieval audience has revealed the impact of *memory* on the interpretation of the work. Rather than taking it as a cue to think about the role a secular ruler would play in the future (as modern scholars have often wrongly assumed) (Holdenried 2006: 13–30), the Tiburtina's medieval scribes and readers repeatedly recalled that the acrostic poem which concludes its narrative was the *Vos inquam* homily from the Christmas liturgy.

---

[27] As is all too rarely acknowledged, a future application (i.e. updating of the regnal list) is also rare in the period *before* 1200, see Holdenried (2006: 20–22 and 41–52).

The significance of this from a temporal perspective is that *Vos inquam* was not about the future: it is an excerpt from a sermon that adduces evidence from Jewish and pagan prophets that Christ really was the Messiah. In other words, it provides non-Christian authority about the authenticity of the Incarnation, a historical event.[28] Thus, a backward-looking approach to the Tiburtina was at least as common as a forward-looking 'prophetic' attitude.

In this regard, it is striking that the text continued to be copied frequently into the fifteenth century, long after the regnal list fossilised around 1200. If, as seems likely, during this last part of its life, the text was still being given some sort of future-facing interpretation, that interpretation cannot have understood the work as a political prophecy because the rulers indicated in the king list as potential Last Emperors were all now long dead. Indeed, as I have demonstrated elsewhere, post-1200, if not before, the surviving manuscript evidence suggests the prophecy had a different forward-looking interpretation (at least sometimes). This second future-oriented interpretation maintained the text's expectation of a Last Judgement but transformed its eschatological message from a universal one into an essentially personal meaning that did not require contemporary political relevance (Holdenried 2006: 93–108).

In the terms I have been discussing in this chapter we can therefore observe a change in the way the text was used, away from a literal interpretation of its historico-prophetic content (the Last Emperor narrative) towards an approach shaped by the broader cultural tradition of reading biblical *Realprophetie* allegorically. In the following, I view this change through the prism of medieval regimes of temporality to explore how this shift may have been influenced by mental habits for handling time that were the natural by-products of the application of certain techniques of exegesis.

There were two chief means for the Tiburtina's audience to come to understand the rules of scriptural interpretation: one was direct study of the Bible, and the other was to learn exegetical techniques from the liturgy. During the Divine Office, extracts from scripture are recited

---

[28]This connection left a significant impact on the text's manuscript tradition in the form of marginal annotations and amendments to the text, see Holdenried (2006: esp. 111–130).

or sung, together with commentaries, that is, with extracts from actual works of exegesis (Boynton 2006: 64–65). Commenting specifically on the lessons of Matins, it has been noted that they

> represent the forms of biblical commentary and theological discourse that monks heard and read most frequently and at greatest length. While hagiographical lessons offered models of behavior for the monks to contemplate and emulate, exegetical lessons provided examples of interpretation and homiletic techniques. (ibid., 67)

In short, those participating in the Divine Office could learn from the liturgy how to approach the interpretation of text and also as a result, for current purposes, how to understand time. Note, too, that the foregoing quotation is about the lessons of Matins, and that the acrostic poem from the Tiburtina was itself one of those very lessons. It would seem natural for participants in the liturgy to associate such a text with the exegetical lessons that also accompanied it in the rest of the Divine Office.

Thus, in addition to study of the Bible, the liturgy could be another important conduit for the transmission of regimes of temporality based on exegetical techniques, in our case the literal/historical technique and the typological/allegorical technique previously discussed with reference to Peter Damian's letter.[29] In the liturgy, the typological/allegorical technique was particularly prominent because the Divine Office presents the faithful with interpretations of the broader meaning of Christ's life and of other sacred events within the history of salvation. As noted earlier, a typological/allegorical reading of a scriptural passage gave an episode some continuing contemporary or future meaning: this produced a multi-layered sense of time which "blur[red] the distinction between the distant past ... and the present experience of liturgical time and commemoration" (Boynton 2006: 37). Different categories of time (past, present, future) thus coexisted in the liturgy without being perceived as necessarily separate or distinct from one another. As noted,

---

[29] Note that here I only consider the liturgy's role in transmitting exegetical works and techniques (and their attendant regimes of temporality). Of course, the liturgy was also itself the subject of exegetical works about the divine office, which thus developed ideas about time. This point lies outside the scope of the present discussion, but see, for example, Czock (2016).

this perception of time is the key ingredient of the synchronous regime of temporality and was not limited to the confines of the Divine Office. As Boynton has illustrated with reference to Gregory of Catino (ca. 1060–1135), a monastic historian and near-contemporary of Peter Damian (d. 1072), these temporal structures of the liturgy (which rested on exegetical techniques) could shape an individual's thinking. This idea opens up new avenues for thinking about the post-1200 phase in the history of the Tiburtina's repurposing. As mentioned, in this phase the regnal list was no longer updated to include contemporary rulers, implying that the Tiburtina's scribes were no longer much interested in this part of the text; the text had ceased to relate to the political future.

How does the emergence of such seeming indifference to the prophecy fit the chronological and synchronous regimes of temporality of the medieval period? A sideways glance at Peter Damian's letter is instructive here. Peter Damian reports a reading of the events in the Old Testament as having "passed away with age" and as serving "no useful purpose whatsoever"; he blames this on ignorance of the rules of allegorical Bible interpretation. Without allegory, there is a sense of discontinuity between past and present. The Tiburtina, of course, is not a scriptural narrative. It is also a descriptive, not allusive, narrative. On no front does it therefore invite an allegorising approach: if anything, it discourages it. The text's only figurative element, the dream of the nine suns which become progressively more bloodstained, is already presented together with its interpretation; that is, the text states that the nine suns represent the nine *generationes* of mankind, cutting short any attempts to discover hidden meanings. The text positively demands a historical/literal approach. Yet this was not how it was approached, at least after about 1200.

Perhaps, as Peter Damian's letter suggests, the literal/historical approach brings with it a sense of discontinuity from the past, or as he puts it, the sense "that this event has now passed away with age, and that today it should have no further interest for us." Conceivably it was this sense of discontinuity that weakened the impulse among the Tiburtina's scribes to connect the past with the present and the future, that is, it weakened the very impulse needed to keep the list of rulers in the Tiburtina up to date. As the surviving Tiburtina manuscripts show, this impulse faded drastically towards the end of the twelfth century.

Why did this happen at that time? One can only speculate, but it is striking that from the late eleventh century there had been a marked resurgence of interest in literal readings of the Bible, and by the mid-twelfth

century the practice had been taken up by the Paris schools.[30] Allowing
for a small time lag so these new ideas could percolate down to local
scribes, this is just the moment when the Tiburtina's regnal list ceased to
be updated. The fact that the fossilisation of the regnal list and new depar-
tures in exegesis occurred very roughly at the same time has never been
noted before. Whether this change in exegetical preference did indeed spill
over into approaches to the Tiburtina remains unknowable, but the possibility that it may have done so is well worth considering and might explain
this, one of the most noteworthy changes in the text's development.

The Tiburtina's textual link with the liturgy also embedded it in
another temporal structure. As Boynton explains, "the structure of the
liturgy ... links widely separate events ..., constructing a perception of
time as multilayered simultaneity rather than linear progression" (2006:
36). In short, we have here the synchronous regime of temporality with
its blurring of the boundaries between past, present, and future. This
framework permitted readers to connect, as present-day individuals, to
the now distant events of Christ's life and His resurrection, which the
Tiburtina describes; it invited readers to reflect on the significance of
these events (especially of Christ's death) for their own salvation. Again,
the Tiburtina's surviving manuscript evidence suggests that readers
adopted precisely such a reading repeatedly over the course of its literary life. They recognised in the Tiburtina the Gospel story of Christ's
birth, death, and resurrection and then, on the basis of habits of mind
fostered by the liturgy's commemoration of Christ, in reading it these
readers reflected on their own chances of salvation on Judgement Day,
dramatically depicted by the Tiburtina's concluding acrostic (Holdenried
2006: 93–108).

Quite naturally when doing this they adopted the synchronous regime
of temporality, and read the Tiburtina on the basis that, as Peter Damian
expressed it in another context, "whatever then occurred visibly, is
adapted to our need by allegorical interpretation, as the age long past is
made to serve us at the present time."

In sum, the way the Tiburtina's prediction of the future was received
by its audience (mostly monastic/clerical) reveals the operation of two
different regimes of temporality, the chronological and the synchronous

---

[30]This point in time involved, for example, scholars such as Andrew of St. Victor
(d. 1175), see van Liere (2014: 130–139).

regimes: both were rooted in the hermeneutic rules of scriptural exegesis but could also be applied to the genre of non-scriptural prophecy. This point is important because Peter Damian's letter is a discussion *about* Old Testament texts whereas the changes to the non-scriptural Sibylla Tiburtina are changes *to the text itself*. Despite this difference, both sets of reactions represent reader responses, albeit in different forms, and show readers applying the same two regimes of temporality (synchronous and chronological) across different types of prophetic material.

This response is significant because in the Middle Ages reading any prophecy involved using a complex set of ideas about the organisation of three levels of time (past, present, and future) and their position in relationship to each other. The material discussed in this chapter suggests that the way these relationships were understood might vary from reader to reader, but only within limited parameters; this is because the normative rules for understanding time in the Middle Ages arose from the exegetical techniques adopted by readers when interpreting a text. Readers thus had only a limited set of options from which to choose. If they chose to adopt the synchronous regime, the prophecy was repurposed by treating it as a metaphor and thus rendering it atemporal, that is, timelessly applicable across past, present, and future. If they chose the chronological regime, on the other hand, then, like the ratchet, they repurposed the prophecy by shunting it forward in time to keep it current, lest, as Peter Damian's querulous persons said, it became redundant. Note that although I refer here to a reader's 'choice', that may not have always been a conscious decision, but rather part of an almost unthinking approach to textual interpretation and everything that flowed from that.[31]

Last, note that the consequence of a reader's choice of regime of temporality sometimes has surprising effects. For example, as I have argued, the chronological regime could render the past separate and detached from the present, deserving only indifference from the reader: this is often considered a 'modern' perspective. Somewhat ironically then, it was precisely the regime of temporality with the most obvious connection to the act of updating prophecy that may have also discouraged such a response.

---

[31] As an example of this, see Fleming (2013: 82).

BIBLIOGRAPHY

Adam, Barbara. 2004. *Time*. Cambridge: Polity Press.

Auerbach, Erich. 1938. Figura. *Archivum Romanicum* 22: 436–489.

———. 1984. Figura. In: *Scenes from the Drama of European Literature*, trans. Ralph Manheim. Theory and History of Literature, vol. 9, 11–76. Minneapolis: University of Minnesota Press.

———. 2014. *Time, History, and Literature: Selected Essays of Erich Auerbach*, ed. James I. Porter, trans. Jane O. Newman. Princeton: Princeton University Press.

Balke, Friedrich, and Hanna Engelmeier. 2016. *Mimesis und Figura: Mit einer Neuausgabe des "Figura"-Aufsatzes von Erich Auerbach*. Paderborn: Wilhelm Fink.

Borst, Arno. 1993 [1990]. *The Ordering of Time: From the Ancient Computus to the Modern Computer*, trans. Andrew Winnard. Chicago: The University of Chicago Press.

Boynton, Susan. 2006. *Shaping a Monastic Identity: Liturgy and History at the Imperial Abbey of Farfa, 1000–1125*. Ithaca: Cornell University Press.

Burke, Peter. 2004. Reflections on the Cultural History of Time. *Viator* 35: 616–626.

Czock, Miriam. 2016. Vergangenheit, Gegenwart und Zukunft—Konstruktion von Zeit zwischen Heilsgeschichte und Offenbarung: Liturgieexegese um 800 bei Hrabanus Maurus, Amalarius von Metz und Walafrid Strabo. In *ZeitenWelten: Zur Verschränkung von Weltdeutung und Zeitwahrnehmung, 750–1350*, ed. Miriam Czock, and Anja Rathmann-Lutz, 113–134. Cologne: Böhlau.

Dales, Richard C. 1990. *Medieval Discussions of the Eternity of the World*. Leiden: Brill.

Fleming, John V. 2013. Prophecy and Exegesis in the Visio Attributed to Joachim of Fiore. In *Joachim of Fiore and the Influence of Inspiration: Essays in Memory of Marjorie E. Reeves (1905–2003)*, ed. Julia Eva Wannenmacher, 75–98. Church, Faith and Culture in the Medieval West. Aldershot: Ashgate.

Gallois, William. 2007. *Time, Religion and History*. Harlow: Pearson Education.

Hartog, François. 2015 [2003]. *Regimes of Historicity: Presentism and Experiences of Time*, trans. Saskia Brown. New York: Columbia University Press.

Holdenried, Anke. 2006. *The Sibyl and Her Scribes: Manuscripts and Interpretation of the Latin Sibylla Tiburtina, c. 1050–1500*. Aldershot: Ashgate.

———. 2014. Many Hands Without Design: The Making of a Medieval Prophetic Text. *The Mediaeval Journal* 4 (1): 23–42.

————. 2016. Christian Moral Decline: A New Context for the Sibylla Tiburtina (Ms Escorial & I.3). In *Peoples of the Apocalypse: Eschatological Beliefs and Political Scenarios*, ed. Wolfram Brandes, Felicitas Schmieder, and Rebekka Voß, 321–336. Millennium Studies 63. Berlin: De Gruyter.

Hunt, Lynn. 2008. *Measuring Time, Making History.* The Natalie Zemon Davis Annual Lectures. Budapest: Central European University Press.

Jerome. 1970. *Sancti Eusebii Hieronymi Epistulae*, vol. 3, ed. Isidorus Hilberg. New York: Johnson.

Jordheim, Helge. 2014. Introduction: Multiple Times and the Work of Synchronization. Forum: Multiple Temporalities. *History and Theory* 53 (4): 498–518.

Koselleck, Reinhardt. 2004 [1979]. *Futures Past: On the Semantics of Historical Time*, trans. Keith Tribe. New York: Columbia University Press.

Lerer, Seth (ed.). 1996. *Literary History and the Challenge of Philology: The Legacy of Erich Auerbach.* Stanford: Stanford University Press.

Munn, Nancy D. 1992. The Cultural Anthropology of Time: A Critical Essay. *Annual Review of Anthropology* 21: 93–123.

Peter Damian. 2005. *Letters 151–180*, trans. Owen J. Blum and Irven M. Resnick. The Fathers of the Church: Mediaeval Continuation. Washington, DC: Catholic University of America Press.

Reindel, Kurt (ed.). 1993. *Die Briefe des Petrus Damiani: Teil 4, Nr. 151–180. Monumenta Germaniae Historica: Die Briefe der deutschen Kaiserzeit*, vol. 4. Munich: Monumenta Germaniae Historica.

Spiegel, Gabrielle. 2016. Structures of Time in Medieval Historiography. *The Medieval History Journal* 19 (1): 21–33.

Stern, Sacha. 2003. *Time and Process in Ancient Judaism.* Oxford: The Littman Library of Jewish Civilization.

van Liere, Frans. 2014. *An Introduction to the Medieval Bible.* Cambridge: Cambridge University Press.

CHAPTER 3

'Wunschzeit' Jerusalem: Rethinking the Distinction Between Time and Space in Medieval Utopias

Christian Hoffarth

'No-Place-Land' and 'Neverland'

The fascination with unknown time that can be found amongst Christians in the Middle Ages mainly derived from two sources. The first source was determinism, because Christians believed that the whole of history followed a divine plan. As a consequence, it became characteristic of medieval thought to think of certain future events as being already fixed in time—at least to a certain degree. This conviction led to a much stronger fascination and open-mindedness towards means of prediction than it would have been the case if people had believed in an 'open' future.[1] The second source of the fascination with unknown time

I owe thanks to Michael A. Conrad and Erik Collins for proofreading this article and for their additional comments.

[1] On medieval prediction and prophecy, see Chaps. 2 (The Old Made New) and 5 (Unknown or Uncertain?) of the present volume.

C. Hoffarth (✉)
University of Duisburg-Essen, Essen, Germany

© The Author(s) 2017
S. Baumbach et al. (eds.), *The Fascination with Unknown Time*,
DOI 10.1007/978-3-319-66438-5_3

in medieval Latin Europe had to do with the perception of the Bible as representing the Word of God. If history was predetermined by God, its course should consequently reveal (and already have revealed) itself in the words of Holy Scripture. Making unknown time known was thus merely a question of the correct method. The general idea of a predetermined future, however, did not lead to a resignation to fate (Schmitt 2000: 11–13). On the contrary and despite common stereotypes, the Middle Ages knew various forms of hope and planning for a better life in this world—not only of the individual but of mankind in general[2]— and indeed there was a way of thinking that can be described as *utopian*. Based on and starting from these observations, this chapter explores the temporal element of medieval utopian thought, asking about its ways of expression, its function, and the epistemological foundations of its inextricable entanglement with spatiality.

A convenient starting point for the following argument is provided by the etymology and development of the word *Utopia* itself: the term is derived from Greek *ou* (not) and *topos* (place) and can be translated literally as "no-place-land." The very word *Utopia*, however, was not known in ancient Greek but was coined as a neologism by the English humanist Thomas More (1478–1535).[3] In 1516, More, under the editorship of Erasmus of Rotterdam (1466/1469–1536), published his famous book about an imaginary ideal society. The first edition was entitled *Libellus vere aureus nec minus salutaris quam festivus de optimo reipublicae statu, deque nova Insula Utopia*, which translates "A truly golden little book, no less beneficial than pleasing, of a republic's best state and of the new island Utopia." Judging from the etymology of the neologism *Utopia*, but also from the fact that More's best state is placed on an island, it is quite clear that the author had in mind an ideal place, located in an undiscovered or imaginary part of his contemporary world. The setting in his own day is corroborated by the fact that More uses the form of an oral travelogue, given by Raphael Hythlodeus, a fictitious companion of Amerigo Vespucci. More himself asserts to have witnessed the presentation of Hythlodeus's report in Antwerp in the company of his fellow humanist Pieter Gillis (c. 1486–1533) (More [Logan et al.] 1995: 43–45).

---

[2] For a first impression and overview on this subject, see Seibt (2001).

[3] On further implications of the word, especially with reference to the homophony *Eutopia* ("good place"), see, e.g., Vieira (2010: 3–5).

Some of the author's coevals, however, apparently did not take the contemporary setting of *Utopia* for granted. Shortly after More's book had been published, French humanist Guillaume Budé (1468–1540), in a (probably fictitious) letter to Englishman Thomas Lupset (c. 1485–1530), suggested that it would be justified to also call More's work *Udepotia*.[4] This additional Greek neologism derives from the adverb *oudepote* (never) and could therefore be translated as "Neverland." In the two subsequent editions of *Utopia*, published in Paris and Basel in 1517 and 1518, Budé's letter was printed as a preface with More's (at least tacit) approval.[5] This way, *Utopia* and *Udepotia* were united between the covers of one single book. Figuratively speaking, one could say that both terms became 'heralds' of the new, ideal society, flanking it side by side.

Hence, in the eyes of sixteenth-century humanists the concepts of "No-place-land" and "Neverland" were naturally linked to each other as two entries to one and the same subject. It seems that for these humanists the crucial element of *U-topia* was not so much the *-topia* part but rather the prefix *u-*; in other words, they were less interested in the notion's spatial aspect but rather in the fact of its nonexistence.[6] Designed as a counter-image to a reality that was considered to be flawed (Elias 1982: 113–133; Nipperdey 1975: 114–115), *Utopia's* persuasiveness mostly depended on its fictionality. Its setting, then, could likewise be rendered as either spatial or temporal, without affecting its nonexistence as its pivotal characteristic.[7] If it could not exist in either present or future, this meant that it could not exist in either space or time.

---

[4] "VTOPIA vero insula, quam etiam VDEPOTIAM appellari audio. ..." An edition and English translation of the letter in Thomas More, *Utopia*, eds. Logan, Adams, and Miller, 1995: 6–19; the quotation ibid., 12. On Budé, see McNeill (1975); on his letter to Lupset, ibid., 50–51.

[5] McNeill points out that More "certainly did not seek to remove it [the letter] from the subsequent editions. He thus implicity [*sic*] approves of Budé's interpretation" (1975: 50). On More's involvement with the editions of 1517 and 1518, see Thomas More, *Utopia*, eds. Logan, Adams, and Miller, 1995: pp. 271–273.

[6] Accordingly, in scholarship on More's work as well as on *Utopia* as a genre of literature or a pattern of thought, the non-existence and impossibility of realisation is often emphasised as a central criterion to conceive of a certain concept as utopian. Cf. Schölderle (2011: 463–465).

[7] Zudeick assumes that to More the spatial aspect had been more important because More had wanted to emphasise the non-existence of his concept (2012: 634). Funke on the contrary claims that Budé's *Udepotia* stresses the factor of infeasibility (1991: 11). However, neither commentator explains how one of the two concepts, "No-place-land"

On the basis of these observations, I want to argue that the blending of unknown times and unknown spaces is a traditional feature of utopian thought, which had already been well established before Thomas More in the Middle Ages. In doing so, I would like to challenge a widespread and popular topos of historic research and especially of medieval studies: in fact, many authors assume that premodern visions of an ideal world could be easily and clearly distinguished as either spatial or temporal.[8] This assumption, however, not only leads to inaccurate ideas about the utopian in the Middle Ages but it also obscures the specifics of the fascination with unknown time in this era.

To revise this assumption, it is first of all necessary to identify its origin and its appeal to medievalists by recalling, in some detail, a number of relevant milestones of the scholarship on medieval utopianism. The necessity to devote a substantial part of the chapter to research tradition is the result of the topic being highly contested. Especially in the twentieth century, it spawned a range of ideologically informed conflicting paradigms.[9] Reviewing previous scholarship will, therefore, not only lay the ground for rethinking the distinction of time and space in medieval Utopias but also provide an insight into the methods later thinkers have used to adapt past futures to their own positions on unknown time.

The following discussion of 'Jerusalem' as a utopian concept will then demonstrate that the temporal and spatial dimensions in medieval utopian thought have the capability of converging, thereby becoming indistinguishable. To achieve this goal, it is necessary to discuss the specific relationship between the earthly and the heavenly Jerusalem with a focus on biblical exegesis and monastic self-understanding. Some of these observations will lead to the main example of an investigation into the convergence of time and space in medieval Utopias: Joachim of Fiore's (ca. 1135–1202) outline of an ideal society in his *Liber figurarum*.

---

or "Neverland," could have appeared, to More, to be 'more unreal' than the other. Rather, More's decision for a spatial depiction of the ideal state might have been from the wish for authenticity that was to be gained by the allegedly non-fictional genre of travelogues.

[8] This idea comes from Alfred Doren (cf. Doren 1927) and has been approved by virtually all contributors to the study of medieval utopian thought. See the discussion below.

[9] This is true not only of Marxist approaches, for example, but also regarding the conflict between medievalists and modernists.

WUNSCHRÄUME AND WUNSCHZEITEN—SCHOLARSHIP ON MEDIEVAL UTOPIAS

For a long time, the Middle Ages had been considered a 'utopian vacuum' by historians and other scholars. It was not before 1927 that Alfred Doren published a pioneering article on the history of utopian thought, in which he treated the period from Antiquity to the nineteenth century, reserving an important part for the Middle Ages. With this article (Doren 1927), Doren finally provided the basis for an interpretation of certain traits of medieval thought as utopian. In addition, Doren also developed a typology that still remains influential today: He distinguishes two main forms of desire for a better world, which he labels *Wunschräume* (wishing spaces) and *Wunschzeiten* (wishing times),[10] assigning the former to Utopias and the latter to a variety of types of chiliasm (1927: 158n1). He further defines *Wunschräume* as an "overcoming of the sensually comprehensible known space by a deliberate pictorial projection of a *Wunschraum* on an imaginary and barely conceivable geographical surface" and *Wunschzeiten* as "ideal prolongations of chronologically recognisable events in the sense of a necessary advancement towards imaginary ends somewhere on the edge of time" (ibid., 1927: 161 [my translation]). With regard to the Middle Ages, Doren assumes that the Christian chiliastic anticipation of an imminent earthly paradise (see Cohn 1993; Töpfer 1964; McGinn 1979a, 1998; Aertsen and Pickavé 2001; Landes et al. 2003) had been the exclusive form of thinking that transcended reality. Thus, Doren concludes that during the centuries before Thomas More there had only been temporal, but not spatial, visions of ideal worlds (1927: 182–183).

In contrast, medievalist František Graus claims that "[t]he modern concept of an ideal society in the 'historical' future is foreign to the Middle Ages" (1967: 8–9). Here he draws on the distinction between earthly and eschatological history that, indeed, was crucial to the medieval understanding of time. I return to this later. To Graus, the chiliastic belief in a restoration of paradise in the earthly future was only an exception to the general setting of medieval Utopias on far-off islands or in a past golden age (ibid., 8). This position seems to be sustainable

---

[10] *Wunschraum* and *Wunschzeit* are German words literally referring to imaginary spaces or times wishfully desired. As there seem to be no fully adequate translations, I will retain the German terms.

only by excluding from the picture—as Graus does—the broad influence of Joachim of Fiore's theology of history that facilitated the perception of history as progress since the thirteenth century.[11] Because Graus's approach thus only covers part of the picture and had no particular impact on later theories of the utopian, it can be left aside for now. In the following, I wish to return to Doren's much more influential considerations.

Following Doren's distinction between Utopias and chiliasms, it might seem that there could not have been any Utopias between Antiquity and Renaissance. If we accept, however, that Doren perceived his *Wunschräume* and *Wunschzeiten* to be dreams of "redemption of a suffering mankind beyond the realm of sensual perception" and "expressions of human yearning, namely not of the yearning for an ideal existence of a single human being but of mankind as such" (Doren 1927: 161 [my translation]), we may conclude that he broadened the general view on utopian concepts beyond boundaries of genres and eras, and, at the same time, laid the foundation for a more differentiated historiography of utopian thought (Oexle 1997: 1346).

Accordingly, Doren's typology of *Wunschräume* and *Wunschzeiten* was well received at the moment of its publication. Karl Mannheim, for instance, confirmed the distinction between chiliasms and Utopias as appropriate "descriptive principles," and even considered Doren's work to be "the best guide for the treatment of the problem from the point of view of cultural history and the history of ideas." As far as his own purposes were concerned, that is, his attempt to determine the function of Utopias in the formation of modern consciousness, however, Mannheim considered the distinction between *Wunschräumen* and *Wunschzeiten* to be "only of indirect value" (1954: 184–185 with n1). His own theory of utopian thought is, in fact, much more inclusive than Doren's, which limited itself to spatial visions of an ideal world. Mannheim on the other hand defines Utopias as "all situationally transcendent ideas (not only wish-projections) which in any way have a transforming effect upon the existing historical-social order" (ibid., 185). In line with this, he

---

[11] As Graus notes, his position is indebted to A.O. Lovejoy's and G. Boas's studies on primitivism (Lovejoy and Boas 1935; Boas 1948). Interestingly enough, however, Boas himself had already noted that there had also been a tradition of anti-primitivism in the Middle Ages and had identified Joachim as one of its main contributors (Boas 1948: 206–216).

unhesitatingly assigns medieval chiliastic *Wunschzeiten* to the category of the utopian (ibid., 190–197).

The Marxist philosopher Ernst Bloch, too, used terms and phrases by Doren without quoting him explicitly,[12] and, similar to Mannheim, supported a very broad definition of utopias. In Bloch's system, utopian thought is understood as an 'anticipating consciousness,' a manifestation of possibilities, and the metaphysical status of the 'Not-Yet' (Zudeick 2012; Siebers 2012). This theory allowed Bloch, too, to subsume medieval ideas of both spatial and temporal 'better worlds' under the category of Utopia.[13]

Mannheim and Bloch have thus established concepts that claimed universal validity. An unfortunate side effect, however, consisted in their conceptual vagueness that entailed a lack of heuristic potential—certainly one of the reasons why both theories have not been well received by later scholars. Instead, Bloch's magnum opus *The Principle of Hope* has often been understood as a sourcebook rather than as providing a feasible theory of utopian thought. Accordingly, Bloch was not considered "a theorist of the utopian but its phenomenologist and encyclopaedist" (Neusüss 1972: 88 [my translation]). It seems that, for similar reasons, his inclusion of medieval concepts into a broad history of utopian ideas had only little impact on later historians.

In the 1960s and 1970s, prominent medievalists joined the debate on the history and development of utopian thought.[14] Although all of them, too, accepted Doren's theory, they also tried to demonstrate that certain medieval *Wunschzeiten* showed characteristic features of utopian thought (as opposed to Doren's categorical distinction between Utopias and chiliasms). In addition, they argued that spatial Utopias had also existed during the Middle Ages. In this context Ferdinand Seibt referred to the concept of monasticism: He argued that the highly formalised communitarian lifestyle of monks in the cloister—an 'island within society'—can

---

[12] See, for instance, Bloch (1972: 217): "… alle die Wunschzeiten und Wunschräume der alten Utopie …"; Bloch (1986: 68): "… auf eine utopische Bildwand projiziert." Cf. Doren (1927: 189n49).

[13] On spatial Utopias in the Middle Ages, see Bloch (1969, 2: 889–909). On the temporal dimension see, for instance, his account on Joachim of Fiore, ibid., 1: 590–598.

[14] The first was Graus (1967), who hardly reflected on the question if the concept of Utopia could be transferred to the Middle Ages at all.

be understood as a utopian *Wunschraum*.[15] Moreover, Seibt underlined that medieval chiliasm could also be interpreted as a form of utopian thinking. Starting from the idea of a "secularisation of chiliasm" (my translation) in the fifteenth century, he stated that, despite their religious basis, chiliastic movements were basically oriented towards an improvement of the status quo (Seibt 2001: 145–193). In a similar way, Otto Gerhard Oexle objected to the notion that the Middle Ages had only been familiar with chronological forms of utopian ideas, that is, ideal unknown times (Oexle 1977). Furthermore, in agreement with Seibt, Oexle strongly argued against the widespread conviction that the term *Utopia* should be restricted to the description of modern phenomena. Seibt's and Oexle's work gave birth to a fruitful but still under-explored field of research on utopian thought in the Middle Ages.[16]

In spite of these groundbreaking studies, there still are—and always have been—political scientists and historians, especially of the Early Modern Period, who insist that utopian thought represents a signature of modernity, thereby denying its existence during the Middle Ages (e.g., Kamlah 1969: 16; Nipperdey 1975: 125; Saage 2001–2003, 1: 48ff.). Most of these authors continue to apply Doren's distinction between chiliasm and Utopia and argue that the former had been so profoundly different from the latter that it could simply not be subsumed under the category of the utopian. Excluding the Middle Ages from the history of utopian thought, such researchers usually assume a "temporalisation of Utopia" (Koselleck 1982) in the eighteenth century. These two positions still irreconcilably oppose each other in current historical debates.[17]

It is not the intention of this chapter to decide once and for all whether there had been utopian thought in the Middle Ages. To a certain degree, this must be considered a pointless discussion anyway, as it primarily depends on the question of definition, whereas the different

---

[15] Cf. Seibt (1969: 563, 2001: 15–16, 2002: 310). The same idea can be found in Frye (1965: 333). More references are given by Oexle (1977: 316n99).

[16] See the bibliography in Hartmann and Röcke (2013a: 10–13).

[17] See, e.g., Saage (2013) who against all other contributors to the same volume on Utopia in the Middle Ages holds the opinion that the utopian is a genuinely modern phenomenon. In one of the most recent contributions to the subject, Lochrie (2016: 1–5) reviews the discussion and, as a medievalist like Oexle and Seibt, concludes that there had been utopian thought before Thomas More. However, she seems to be largely unaware of the broad German research tradition.

definitions of the utopian advocated by scholars concerned with the question of the utopian in the Middle Ages seem in turn to be largely based on ideological prejudices about the possibility of utopian thought in this very era.[18] After all, the whole debate thus boils down to a fight between circular references. The assertion that the term *utopian* would lose its distinctness by also assigning it to medieval phenomena (Saage 2001–2003, 1: 66) is, however, untenable. Tomas Tomasek, for example, gives a definition of the utopian that offers a strong heuristic distinctiveness without being restricted to certain epochs. In reference to Neusüss and Horkheimer, he describes utopian thought as a "'dream of the true and righteous order of life' ... that pursues a double goal: a 'criticism of that what is' and a 'depiction of that what should be,'" and explains further that, following this definition, not "every idyll or every ideal is grasped as a Utopia but only such a concept in which, on the basis of dissatisfaction with the existent, an optimal order of life is projected" (Tomasek 2001/2002: 184–185 [my translation]). From a medievalist perspective, this unprejudiced but sharp account is a very convenient starting point for investigations into the characteristics of certain expressions of a longing for a better life.

Regarding the question of the role of unknown time in medieval utopian thought, the state of research poses yet another problem: Although Seibt, Oexle, and others modified Doren's theory by pointing out that chiliasm had not been the only medieval form of thinking that transcended reality, they did not question the distinction between ideal places and ideal times.[19] It seems that they somehow felt obligated to keep Doren's categories, as it was this typology that had provided them with an argument to refute the assumption that the Middle Ages were a utopian vacuum in the first place (Hartmann 2010: 1402). The price they had to pay for proceeding in such a way was to accept the simplifying distinction between *Wunschzeiten* and *Wunschräumen*. It

---

[18]Mannheim comments: "The very attempt to determine the meaning of the concept 'utopia' shows to what extent every definition in historical thinking depends necessarily upon one's perspective. ... The very way in which a concept is defined and the nuance in which it is employed already embody to a certain degree a prejudgment concerning the outcome of the chain of ideas built upon it" (1954: 177).

[19]See Seibt (1969: 556), Oexle (1977: 302–305). In another paper, Oexle even borrows Doren's title "Wunschräume und Wunschzeiten," thereby demonstrating that he endorses these categories, see Oexle (1994).

was only for the modern era that Doren had been able to find examples of occasional convergence of the two forms.[20] More recent research locates the beginnings of such a convergence in the eighteenth century (Koselleck 1982; Saage 2000: 103; Vondung 2000: 31).

So far as I can see, medievalists by and large have not yet questioned the distinction between *Wunschzeiten* and *Wunschräumen*.[21] However, some of the authors and ideas frequently discussed in research on medieval utopianism do not seem to fit Doren's categorisations. This distinction is especially true for the multitude of concepts which in one way or another refer to "Jerusalem" as a utopian archetype.

WUNSCHZEIT JERUSALEM

Oexle's main example of a spatial Utopia in the Middle Ages is a crusade treatise by the French jurist Pierre Dubois (d. ca. 1321) (Oexle 1977: 320–339). In this text, entitled *De recuperatione Terre sancte*, Dubois, differing from other contemporary authors of crusade tracts, presented not so much a plan for reconquering the Holy Land but a draft for building and organising an ideal society in these parts after the successful (re-)conquest (Oexle 1977: 324–325, 335–336; Rexroth 2008: 313).[22] Although this ideal state is, obviously enough, situated in an unknown future, Oexle, in accordance with Doren's terminology, primarily describes it as a *Wunschraum*, because Dubois's concept aims at perfecting society through establishing a model settlement within a geographical area that is precisely defined and confined (Oexle 1977: 336).[23] Dubois did not have "imaginary ends somewhere on the edge of time" in mind, as Doren had put it; instead, he thought of a near future that could and should become real (Brandt 1956: 37, 43). Moreover, the author does not ascribe any eschatological meaning to Jerusalem as the centre of the Holy Land (Oexle 1977: 329). In this respect, Dubois

---

[20] For example, in H.G. Wells, see Doren (1927: 204n68).

[21] Quite on the contrary: The distinction is, for example, explicitly accepted in Hartmann (2010), Hartmann and Röcke (2013b), Stock (2013), and Tomasek (2001/2002), who adds the third category of the *Wunschmensch* (a compound of *Wunsch* 'wish' and *Mensch* 'human').

[22] On late medieval crusade projects in general, see Paviot (2014).

[23] According to Oexle, in contrast to a *Wunschzeit*, a *Wunschraum* is particularly defined by its clear demarcation, cf. Oexle (1994: 38–39).

once more breaks with tradition, because, as Oexle notes, "the idea that Jerusalem's liberation had a certain place in history approaching its end; that the king moving into Jerusalem would induce the thousand-year-long empire of peace; that Jerusalem was the centre of the world, the place in which the peoples would gather, and the place of the Last Judgement—all these ideas belonged to the intellectual premises of the crusade movement" (Oexle 1977: 329 [my translation]).[24] Although there certainly is a temporal dimension in Dubois' outline of an ideal society, he does not, however, align himself with medieval habits of identifying the earthly Jerusalem as a figuration of the heavenly one.[25] To explore this particular aspect in more detail, I now focus on more general medieval approaches to Jerusalem.

Generally speaking, the eschatological interpretation of temporal progress was crucial to medieval Christian thought. Put simply, people in the Middle Ages were certain that earthly existence—not only of individuals but of all mankind—would come to an end; moreover, despite awareness of a long past, they continually believed this end to be near.[26] The events that were to be expected after the end of earthly history and time could be gleaned from the Bible, particularly from the last two chapters of its concluding book, that is, the description of the New Jerusalem in the Revelation of John. In these passages a Holy City of God is presented: a restoration of paradise, as it were, where there will be fruits in abundance, no more sickness, and "no more curse," that is to say, the covenant between God and man will be renewed.[27]

Yet, Jerusalem was more than this 'otherworld' that was to come about at the end of times: the earthly city of Jerusalem had already been the place of Christ's passion—where the son of God had died on

---

[24] On the eschatological meaning of Jerusalem in the idea of crusade, see Auffarth (2002: 73–122), Auffarth (1993).

[25] On the relationship between the earthly and the heavenly Jerusalem in medieval thought, see Causse (1947), Konrad (1965), Kühnel (1987), Stroumsa (1999), Renna (2002), Augustyn (2008).

[26] See Landes et al. (2003), Aertsen and Pickavé (2001), Cohn (1993), McGinn (1979a, 1998), Töpfer (1964).

[27] The description of heavenly Jerusalem in the Book of Revelation uses motifs taken from descriptions of paradise in Genesis, see Gn 2:9, and comp. Rv 22:2, 14 (tree of life); Gn 2:10, and comp. Rv 22:1 (river of life); Gn 2:11–12, and comp. Rv 21:18–21 (three precious materials).

the cross for the salvation of mankind—and of his resurrection, thus becoming the cradle of Christianity.[28] Therefore, in one way or another, Jerusalem had always been an object of desire. But because the concrete and material city in the Middle East had been out of reach and inaccessible for the majority of Christians during most of the Middle Ages, this desire had mostly focused on the heavenly Jerusalem.[29] As a consequence, Jerusalem was not only perceived as an ideal place but perhaps even more as an ideal time—or, to be exact: an ideal eternity.

This form of perception had been made possible, first and foremost, by the influence of biblical exegesis on almost all patterns of medieval Christian thought.[30] One of the main ideas applied to interpret the Bible was that of the fourfold sense of Scripture,[31] according to which every utterance in the Bible has four different meanings: (1) the 'literal' or 'historical' sense; (2) the 'allegorical' sense, by which the words were understood as metaphors or symbols enabling the reader to adopt their meaning to new circumstances; (3) the 'moral' or 'tropological' sense, according to which the Bible was read as a collection of moral rules. Complying with these rules was considered a precondition of individual salvation. Finally, (4) there is the 'eschatological' or 'anagogical' sense, through which readers could find within the Bible the objects of Christian hope.[32]

---

[28] In early Christianity, after it had found a new centre in the city of Rome, there was an ambivalent attitude towards the earthly Jerusalem, for it was the place where Christ had been murdered. Cf. Stroumsa (1999: 34–36), Augustyn (2008: 99).

[29] This had its roots already in the Bible. In Galatians 4:21–31, for instance, the Jerusalem 'that is above' is superordinated to the earthly one. Another tactic to overcome the distance of the earthly Jerusalem was its reconstruction at other places, a *translatio Hierosolymae*. Cf. Auffarth (1993: 101–104), Stroumsa (1999), Jaspert (2001), Bernet (2007).

[30] See de Lubac (1998–2009, vol. 1: xix): "It [premodern biblical exegesis] sets up an often subtle dialectic of before and after. It defines the relationship between historical reality and spiritual reality, between society and the individual, between time and eternity. ... It organizes all of revelation around a concrete center, which is fixed in time and space by the Cross of Jesus Christ. ... It is the principal form that the Christian synthesis had for a long time been shaped by." See also Holdenried in this volume, pp. 33–37.

[31] See de Lubac (1998–2009).

[32] Summarised in a mnemonic distich, probably going back to the Dominican Augustine of Dacia (d. ca. 1285): "Littera gesta docet, quid credas allegoria / moralis quid agas, quo tendas anagogia." On the origins, see de Lubac (1998–2009, vol. 1: 1–2).

Each of the four senses is linked to a specific dimension of time (Barney 1989: 180): the historical to the past, the allegorical to the present, the moral to the individual future, and the eschatological to the collective future or to eternity. Moreover, at least three of these four senses, namely the historical, the moral, and the eschatological ones, are somehow connected to instances of 'unknown time.' Surely, the historical reading of Scripture in medieval perception allowed for learning about human history, which, therefore, could be 'known'—in one sense of the word. The past *condition* of humanity (e.g., the perfection in the state of innocence or in the early church), however, remained alien to later Christians. So, insofar as the realm of experience is concerned, the unknown past inevitably stayed unknown to medieval readers of the Bible. In contrast, the moral sense was directed at one's experiences in a future unknown time and was supposed to help make them controllable. In accordance with Christ's precept to "lay up ... treasures in heaven" (Mt 6:20), leading a moral life during earthly existence was thought to guarantee the individual's redemption in the afterworld. More than the historical and the moral sense, however, the eschatological sense aroused fascination with unknown time. According to Revelation, before the Last Judgment and the installation of the New Jerusalem, a millennial kingdom of peace on earth was to be expected (Rv 20).[33] The eschatological reading of the Bible allowed for an understanding of both God's words and the collectively experienced history as signs of this kingdom of peace. Biblical exegesis thus promised insights into the collective fate of humankind. At least since Joachim of Fiore, it had become common to include, in the picture of God's future empire on Earth, Revelation's depiction of the heavenly Jerusalem, despite it being—technically speaking—located not in the earthly future but in eternity.

The links drawn by medieval writers between the four senses and certain dimensions of time can be vividly illustrated by the most famous medieval textbook example of the fourfold sense: Jerusalem. According to John Cassian (d. ca. 435), Jerusalem, in the Bible, first of all, refers to the city in the Middle East (historical sense), second, to the church (allegorical sense), third, to the heavenly Jerusalem (eschatological

---

[33] On the social and political significance of this idea in the late Middle Ages, see Töpfer (1964).

sense), and, finally, to the human soul (moral sense).[34] Because this example became virtually omnipresent in medieval scholarship, it can easily be assumed that whenever medieval theologians came across the word *Jerusalem*, they would inevitably have recalled all four senses, thereby automatically blending Jerusalem's earthly incarnation with the idea of the heavenly one.[35]

To further explore the hermeneutic mechanisms leading to the convergence of time and space, it is most helpful to look into the exegetical work of the English monk Bede, called the Venerable (672/3–735), which was among the most widespread works during the Middle Ages. In his Commentary on the Song of Songs, written before 716 (Brown 2009: 14), Bede recites the classic example of the fourfold sense of Jerusalem.[36] Then, in his second Commentary on the Acts of the Apostles, composed between 725 and 731 (Brown 2009: 14), he applies it in a very distinctive way. Regarding the life of the first Christian community in Jerusalem, he explains:

> Those who live in such a way that they have all things in common in the Lord are rightly ... called monks. Indeed, this life is the happier the more it withdraws from the other behaviours of this world and the more it imitates in the present age the state of the future age where all goods are common to all the blessed ...; and since there the highest grace of peace and safety is prevailing, the city, in which a type of this life has preceded beforehand, has justly been called Jerusalem, that is, vision of peace. ... (Bede, *Expositio*, ed. Laistner, 1939: 113 [my translation])[37]

---

[34] John Cassian, *Collationes*, ed. Petschenig, 1886: 405: "igitur praedicta quattuor figurae in unum ita, si uolumus, confluunt, ut una atque eadem Hierusalem quadrifarie possit intellegi: secundum historiam ciuitas Iudaeorum, secundum allegoriam ecclesia Christi, secundum anagogen ciuitas dei illa caelestis, quae est mater omnium nostrum, secundum tropologiam anima hominis ...".

[35] Cf. de Lubac (1998–2009, vol. 2: 199–201), with many examples.

[36] Bede, *In Cantica Canticorum*, ed. Hurst, 1983: 260: "Lauda Hierusalem dominum, quod iuxta litteram quidem ciues urbis ipsius in qua templum Dei erat ad laudes ei dicendas hortatur; at uero iuxta allegoriam Hierusalem ecclesia Christi est toto orbe diffusa; item iuxta tropologiam, id est moralem sensum, anima quaeque sancta Hierusalem recte uocatur; item iuxta anagogen, id est intellegentiam ad superiora ducentem, Hierusalem habitatio est patriae caelestis quae ex angelis sanctis et hominibus constat".

[37] "... qui ergo ita uiuunt ut sint eis omnia communia in domino ... κοινοβιῶται uocantur. Quae nimirum uita tanto ceteris saeculi huius conuersationibus felicior est, quanto statum futuri saeculi etiam in praesenti imitatur, ubi sunt omnia omnibus communia bona

At least two important aspects of the medieval idea of Jerusalem are reflected in this passage. First, it demonstrates that the earthly Jerusalem was indeed seen as an instance of the heavenly one and, moreover, as a foreshadowing of a future ideal society. By referring to the common medieval translation of *Jerusalem* as 'vision of peace,' Bede testifies that this idea is even older than his own work.[38] Second, the cited passage highlights an understanding of monastic lifestyle as an earthly precursor of the future New Jerusalem.

In the centuries after Bede, the notion of the cloister as an earthly incarnation of the heavenly Jerusalem became very popular, especially with the order of Cistercians. Bernard of Clairvaux (1090–1153), the most influential Cistercian theologian, introduced the idea into Cistercian identity (Raedts 1994). In a letter to Alexander, Bishop of Lincoln, in 1129, Bernard reports on a canon by the name of Philip whom Alexander had sent on a pilgrimage to Jerusalem. On his way to the Holy Land, Philip had stopped at Bernard's cloister in Clairvaux and decided to stay there instead of continuing his pilgrimage. Bernard explains:

> I write to tell you that your Philip has found a shortcut to Jerusalem and has arrived there very quickly. ... He is no longer an inquisitive onlooker, but a devout inhabitant and an enrolled citizen of Jerusalem; but not of that earthly Jerusalem to which Mount Sinai in Arabia is joined, and which is in bondage with her children, but of that free Jerusalem which is above and the mother of us all. And this, if you want to know, is Clairvaux. She is the Jerusalem united to the one in heaven by whole-hearted devotion, by conformity of life, and by certain spiritual affinity. Here, so Philip promises himself, will be his rest for ever and ever. He has chosen to dwell here because he has found, not yet to be sure the fulness of vision, but certainly the hope of that true peace. ... (Bernard, "Epistula 64," *Sancti Bernardi,* eds. Leclercq and Rochais, 1974: 157–158 [trans. James 1952: 91])[39]

---

beatis ...; et quia ibi summa pacis et securitatis gratia regnat, recte ciuitas in qua huius uitae typus praecessit Hierusalem, id est uisio pacis, dicta est".

[38] See Isidore, *Etymologiae,* ed. Lindsay, 1911, vol. 1: VIII.i.6: *Pro futura vero patriae pace Hierusalem vocatur. Nam Hierusalem pacis visio interpretatur.* Cf. also Augustine, *De civitate Dei,* XIX.xi.

[39] "Philippus vester, volens proficisci Ierosolymam, compendium viae invenit, et cito pervenit quo volebat. ... Factus est ergo non curiosus tantum spectator, sed devotus habitator et civis conscriptus Ierusalem, non autem terrenae huius, cui Arabiae mons Sina coniunctus

As this passage shows, Jerusalem to Bernard, too, had the quality of an eternal idea to which different instances are linked in a universal-instance relationship. To him, however, the cloister seems to have been a more perfect instance of Jerusalem than the city in the Middle East. Figuratively speaking, it provided a link between earth and heaven, lying halfway between this world and the otherworld. Speaking in exegetical terms, the monastery is therefore situated between a moral and an escha-tological sense, between individual and collective future, between future and eternity.

It should have become clear by now that Jerusalem had always been a multidimensional idea of fulfilment, encompassing both temporal and spatial qualities. To what high degree it was possible for *Wunschzeiten* and *Wunschräume* to actually blend together with recourse to 'Jerusalem' can be observed most impressively in a classic example in the scholarship on medieval Utopias: a vision of earthly future by the famous Calabrian prophet Joachim of Fiore.[40]

Joachim had been an admirer of Bernhard of Clairvaux[41] and, for some time, was a Cistercian monk himself. He is particularly known for his unique and very influential theology of history. Put simply, he believed that earthly history unfolds in three increasingly perfect stages, each stage corresponding to one of the personas of the Holy Trinity. According to his writings, after the ages of the Father and the Son, as evidenced by the Old Testament and the New Testament, respectively, a third age would begin around the year 1260: the age of the Holy Spirit. In this last age—"a quasi-millennial era of contemplative perfection on earth" (McGinn 2001: 92)—the monastic form of life would become the dominating condition of all human society (Riedl 2004: 246) (Fig. 3.1).

---

est, quae servit vum filiis suis, sed liberae illius, quae est sursum mater nostra. Et si vultis scire, Claravallis est. Ipsa est Ierusalem, ei quae in caelis est, tota mentis devotione, et con-servationis imitatione, et cognatione quadam spiritus sociata. Haec requies illius, sicut ipse promittit, in saeculum saeculi: elegit eam in habitationem sibi quod apud eam sit, etsi non-dum visio, certe exspectatio verae pacis. ..." On Bernard's letter, see Voigts (2014: 112–114), Pranger (1994: 32ff).

[40] There is a vast literature on Joachim. For general accounts, see McGinn (1985), Reeves (1976), Grundmann (1927), Grundmann (1950), Riedl (2004).

[41] On Bernard's influence on Joachim, see McGinn (1992).

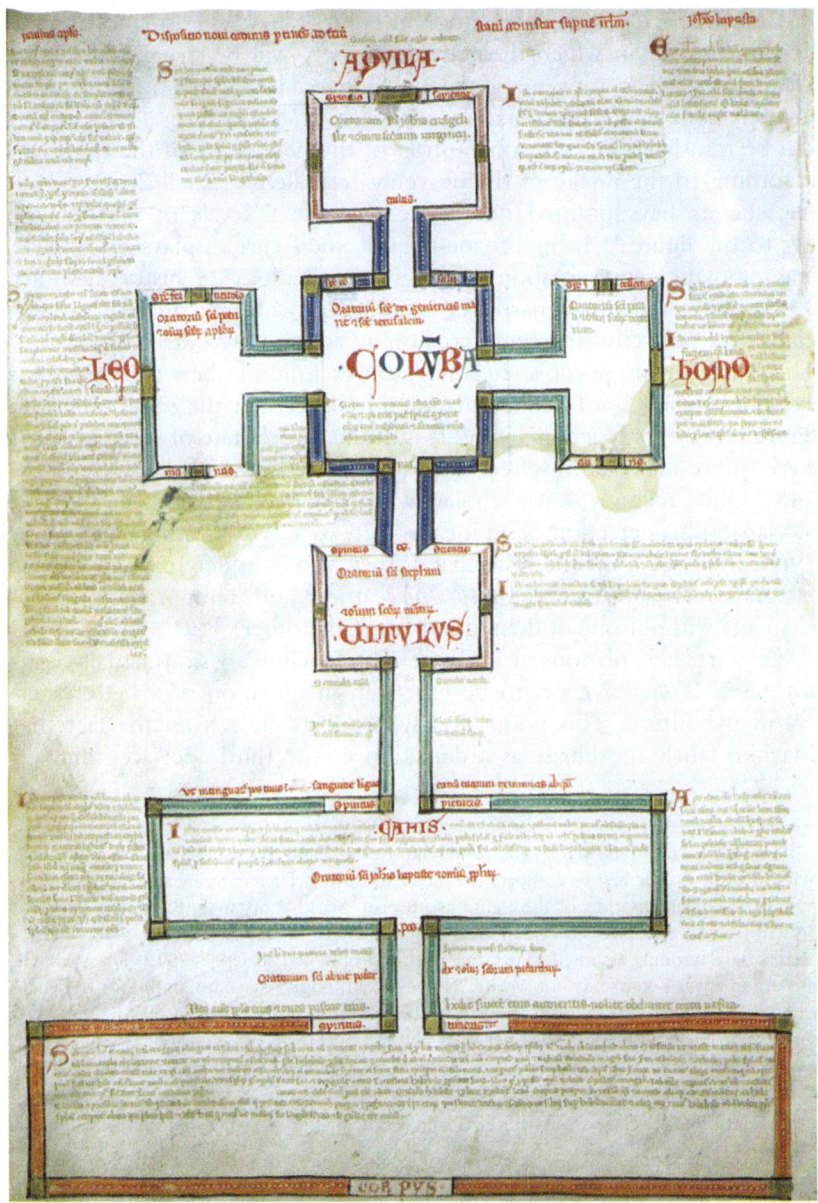

Fig. 3.1 Joachim of Fiore's *Dispositio novi ordinis pertinens ad tertium statum ad instar supernae Jerusalem* from his Liber figurarum (late twelfth century). The figure depicts Joachim's vision of a perfectly organized society, defining it as ideal time and ideal space at once (From Corpus Christi College, MS 255a, fol. 17r. By permission of the President and Fellows of Corpus Christi College, Oxford)

In his *Liber figurarum*, the Book of Figures, Joachim provides an outline of the way in which the order of society will be structured in this ideal time.[42] The caption of the drawing reads "dispositio novi ordinis pertinens ad tertium statum ad instar supernae Jerusalem," which can be translated as "the constitution of the new order of the third age according to the image of the heavenly Jerusalem" (Riedl 2012: 58).[43] As scholars have pointed out, there are several levels of understanding to the figure.[44] In its broadest sense, the figure displays a complete diagrammatic representation of Joachim's theology of history (Müller 1989: 298–304). In a narrower sense, it provides a concrete ground plan for a perfectly structured coenobitic community (Riedl 2012: 59). The image's main purpose, however, can be found in between these two extremes: being both diagrammatic and concrete at the same time, the figure represents Joachim's insights into the future state of society, which were the result of an inspired exegesis (Reeves and Hirsch-Reich 1972: 248). Quite tellingly, most scholars who have dealt with this figure tend to classify this depiction of an ideal society as a *Wunschraum* (e.g., Seibt 1969: 556–557); on the other hand, they also consider Joachim a chiliast. As a matter of fact, the *dispositio* features both temporal and spatial elements without one of them dominating the other.

To start, it is obvious at a glance that Joachim's vision actually constitutes a *Wunschzeit*, or, to be exact, an imagination of a better society in the future. This point already becomes clear from the fact that Joachim labels the image as a depiction of the third age: according to

---

[42] Joachim, *Liber figurarum*, eds. Tondelli, Reeves, and Hirsch-Reich, 1953, Fig. 12. Seibt (1980) was the first to compare the figure to More's *Utopia*. Some scholars, however, refuse the characterisation of the figure as utopian. See, for instance, Riedl: "It does not present us with a utopian society, an ideal with no parallel in empirical reality, as some interpreters have wrongly assumed, but rather with a prophesied society" (2012: 58). As such, this argument, of course, is not wrong. However, although I have no ambitions to decide whether Joachim's disposition is utopian in the full sense of the word, it may be remarked that the problem with Riedl's verdict is, as always, a problem of definition. It seems to be based solely on the assumption that the utopian was defined by fictionality or impossibility of realisation. But see Elias (1982: 144–149).

[43] A full English translation of the text accompanying the picture can be found in McGinn (1979b: 142–148).

[44] In-depth studies of the image can be found in Grundmann (1950: 85–115), Reeves and Hirsch-Reich (1972: 232–248), Thompson (1982), Riedl (2004: 314–334, 2012), Seibt (1969, 1980).

his doctrine, the third age was, after all, and without any doubt, located in the earthly future (Grundmann 1927: 153). Thus, the explanatory description accompanying the drawing was consistently written in the future tense. Yet the reference to the heavenly Jerusalem, from which the depiction borrows its symbolism as well as many architectural details (Riedl 2012: 61–62), should not lead to the conclusion that Joachim would have located his model settlement in eternity rather than in earthly time. He does not depict an Eden-like 'other-world' in which man would be freed from all burdens of life,[45] but a this-worldly order in which the resources of man and earth are used in the best imaginable way.[46] The symmetry of heavenly Jerusalem and the ideal monastery on Earth fully corresponds with the Cistercian tradition and with Joachim's symbolist notion of *Heilsgeschichte*. To ask whether the *dispositio* can be understood as Joachim's idea of the social order in the final stage of history (Reeves and Hirsch-Reich 1972: 245; Riedl 2012: 60n5) or within a pre-stage located between the present and the third age (Grundmann 1950: 111; Thompson 1982: 204) will change nothing about the fact that it was supposed to become real in the future.

And yet, *Liber figurarum* XII can also be considered to represent a *Wunschraum*. The draft's many architectural details imply that Joachim indeed laid out his concept of an ideal monastery complex in this image

---

[45] In Joachim's future society there is infirmity and weakness (I follow the text as given by Grundmann 1950: 116–121): "Sub isto oratorio erunt senes et delicati fratres, qui forte languescente stomacho non possunt ad plenum austeritatem regule in jejuniis sustinere …" (ibid., 117); "… jenunabunt in pane et aqua, excepta causa infirmitas …" (ibid., 118); "Isti jejunabunt omni tempore excepta causa infirmitas. … Si quis autem illorum ita infirmatus fuerit stomacho, ut non possit constitutum sustinere jejunium, transferatur ad oratorium senum …" (ibid., 119); : "Apud istos erit hospitale extra ambitum curtis eorum, in quo erunt lecti strati et cetera necessaria ad usum hospitum sanorum vel infirmorum …" (ibid., 120); "… excepta causa infirmitatis" (ibid., 121).

[46] Joachim uses the metaphor of the body and its members' different spiritual gifts from 1 Cor. 12 to explain how everyone will contribute to the community according to their abilities. See Grundmann (1950: 116). Concerning the laypeople, he states: "Unusquisque ergo operabitur de arte sua; et singule artes vel artifices habebunt prepositos suos. … Mulieres quoque honeste et probate operabuntur lanam ad opus pauperum Christii. … Isti dabunt decimas clericis omnium que possident ad sustentationem pauperum et peregrinorum, set etiam puerorum, qui student doctrine, ea scilicet ratione, ut si hii superhabundaverint et aliqui aliorum minus habebunt, ad arbitrium patris spiritalis accipiatur ab eis, qui plus habent, et dentur his qui minus, ut nullus sit indigens inter eos, et hoc generale erit omnium" (ibid., 121).

and the accompanying textual explanation (Reeves and Hirsch-Reich 1972: 244; Grundmann 1950: 91; Riedl 2004: 314). But this monastery was different: in accordance with the unfolding of monastic life across all layers of society in the third age—laypeople, clerics, and monks—members of all these groups were supposed to inhabit this place, living in a perfect relationship of give and take (Riedl 2012: 64). The serious architectural intention of the figure is highlighted by clear indications as to which part of society should live in which of the seven oratories shown in the image (Grundmann 1950: 92–94). Furthermore, it gives precise instructions about the distance that should be kept between the three main divisions, which housed, in turn, monks, clerics, and laypeople.[47] Joachim also gives instructions on the location and function of buildings,[48] and, concerning a hospice located in the realm of the clerics, he notes that "[i]t will have its own means of support according to its location and type of countryside, both with regards to livestock and to agriculture."[49]

As this cursory analysis already demonstrates, in the thought of the Calabrian abbot Jerusalem was not merely an ideal time or an ideal place, but both at the same time. In other words, to Joachim, Jerusalem became the fixed point in God's creation where time and space coalesced indistinguishably. By imagining an ideal society in the form of monastic life, structured after the model of the New Jerusalem, Joachim applied traditional elements of biblical exegesis and monastic self-esteem, elevating them to the highest exaltation.

---

[47] Grundmann (1950): "Inter hoc monasterium et locum clericorum debet interesse spatium quasi miliarum trium" (119); "Inter hec duo oratoria [of clerics and of laypeople] debet interesse spacium quasi stadiorum trium" (ibid., 120).

[48] Grundmann (1950): "Habebunt autem singuli singulas cellas, ad quas possint cito ingredi, quando volunt orare: non tamen ubi quisque voluerit, sed juxta claustrum ..." (119); "Ad oratorium tamen suum sororem mulierem non recipient; sed ad proprium laycorum oratorium ingredientur diebus festis et celebrabunt apud eos divina officia" (ibid., 120).

[49] Grundmann (1950): "... et habebit proprios redditus secundum loci positionem et qualitatem patrie tam de animalibus quam de agricultura" (120). Trans. McGinn (1979b: 147).

CONCLUSION: UNKNOWN TIME AND THE TRANSCENDENCE
OF REALITY

When it comes to considering the unknown, it seems that, in the Latin Middle Ages, the dimensions of time and space were intimately related. Before the scientific disenchantment of the world, before the formulation of universal laws of nature and the broad exploration and mapping of Earth setting limits to the thinkable (Koselleck 1982: 2–3), unknown spaces could exert fascination similarly to unknown times. Both unknown time and unknown space served as projection screens of fantasy, allowing for the imagination of a better life.

Nonetheless, in the Christian Middle Ages, wishes for a better society and for an ideal order could evolve much more freely within unknown time than within unknown space. That is to say, the Christian dogma of the Fall of Man led to the present being considered as deficient with the deficiencies being regarded as justified punishment for the original sin (Graus 1967: 4). Thus, an overcoming or suspension of these deficiencies in the present was merely imaginable. As in *known space*—with the cloister being interpreted as an earthly reflection of the heavenly Jerusalem—there could only be islands of perfection but certainly no universal perfect community: what could be imagined in *unknown space*, similarly, were only approximations to the ideal. *Unknown time*, on the other hand, allowed for the transcendence of the possible beyond the present deficiencies induced by original sin. In the past—in the state of innocence before the Fall—as well as in the future—in an imagined state of restoration through God—an ideal order was not only possible but rather an unquestionable reality, according to the Bible.

For reasons of its abstractness, however, unknown time was, of course, much harder to picture than unknown space. To conceive an idea of unknown time and to impart it to others, one inevitably had to draw on existing semiotic systems. For this purpose, Jerusalem, which according to the Revelation was the ultimate destination of salvation history, provided the most obvious resource. Consequently, the Bible's depiction of heavenly Jerusalem was applied to both temporal and spatial approaches to the ideal world. This transfer was possible because of certain specifics of medieval epistemology that attributed little to no significance to sensory cognition, instead employing the method of exegesis as the main tool of perception. Exegesis allowed for reading texts in different ways and thus to interpret them either in a spatial or temporal framework, depending on the reader's specific interest. As a result, different inter-pretations could exist simultaneously, which, under other hermeneutic premises, would have contradicted each other. Because the Bible was perceived as the revelation of God's eternal will, it was virtually detached from time and space and thus bore the potential to bring forth mutable time–space relationships that were likewise detached from the temporal and spatial limitations of human nature.[50]

As far as expression was concerned, the fascination with unknown time in the Middle Ages was thus to some degree dependent on spatial semiotics, because only one's own realm of experience allowed for the imagination of the unknown. Joachim of Fiore could not depict the ideal of the third age as anything other than a cloister as there were no other systems of representation available to him.[51] In other words, the desire to visualise and realise unknown time naturally led to a convergence of *Wunschzeit* and *Wunschraum*. Consequently, it becomes clear that medi-eval Utopias cannot always be definitely classified as ideal times *or* ideal places in the way Alfred Doren had suggested. On the contrary, they often encompass qualities of both time and space, or, more concretely, are *Wunschzeit* and *Wunschraum* simultaneously. That being said, the medieval fascination with the temporal element of the utopian—that is, unknown time—lay in the fact that, in contrast to unknown space alone, it allowed for the transcendence of the realm of experience, thereby ena-bling conceptualisation of a human community in which the deficiencies of the present were overcome or, as regarding weakness and infirmity in Joachim's concept, were no longer perceived as deficiencies. In the reli-gious worldview of the Middle Ages, unknown time opened a horizon of utopian possibilities that otherwise would have remained out of reach.

BIBLIOGRAPHY

Aertsen, Jan A., and Martin Pickavé (eds.). 2001. *Ende und Vollendung: Eschatologische Perspektiven im Mittelalter*. Berlin: De Gruyter.

Auffarth, Christoph. 1993. Himmlisches und irdisches Jerusalem: Ein reli-gionswissenschaftlicher Versuch zur 'Kreuzzugseschatologie'. *Zeitschrift für Religionswissenschaft* 1 (25–49): 91–118.

---

[50] See also Holdenried in this volume, pp. 24–25.

[51] Graus states: "It is innate in man to want to strive after an ideal state and to dream one up, and, quite naturally, *his own* concepts and the world *he* lives in provide his starting point" (1967: 5).

Auffarth, Christoph. 2002. *Irdische Wege und himmlischer Lohn: Kreuzzug, Jerusalem und Fegefeuer in religionswissenschaftlicher Perspektive*. Göttingen: Vandenhoeck & Ruprecht.

Augustyn, Wolfgang. 2008. Dantes Paradiso und die Bildtradition zum Himmlischen Jerusalem. *Deutsches Dante-Jahrbuch* 83: 93–113.

Barney, Stephen. 1989. Allegory. In *Dictionary of the Middle Ages*, vol. 1, ed. Joseph R. Strayer, 178–188. New York: Scribner.

Bede. 1939. *Expositio actuum apostolorum et retractatio*, ed. M.L.W. Laistner. Cambridge, MA: Medieval Academy of America.

———. 1983. *Cantica Canticorum, libri VI*, ed. D. Hurst, CCSL 119B. Turnhout: Brepols.

Bernard of Clairvaux. 1974. Epistula 64. In *Sancti Bernardi opera*, vol. 7, eds. Jean Leclercq and Henri Rochais, 157–158. Rome: Editiones Cistercienses.

Bernet, Claus Bernet. 2007. Das Himmlische Jerusalem im Mittelalter: Mikrohistorische Idealvorstellung und utopischer Umsetzungsversuch. *Mediaevistik* 20 (1): 9–35.

Bloch, Ernst. 1969. *Das Prinzip Hoffnung*, 3 vols. Frankfurt am Main: Suhrkamp.

———. 1972. Zur Originalgeschichte des Dritten Reiches. In *Utopie: Begriff und Phänomen des Utopischen*, 2nd ed., ed. Arnhelm Neusüss, 193–218. Neuwied: Hermann Luchterhand.

———. 1986. *Freiheit und Ordnung: Abriß der Sozialutopien*. Frankfurt am Main: Suhrkamp.

Boas, George. 1948. *Essays on Primitivism and Related Ideas in the Middle Ages*. Baltimore: John Hopkins Press.

Brandt, Walther I. 1956. Introduction to *Pierre Dubois: The Recovery of the Holy Land*, ed. Walther I. Brandt, 3–65. New York: Columbia University Press.

Brown, George Hardin. 2009. *A Companion to Bede*. Woodbridge: Boydell.

Cassian, John. 1886. *Collationes*, ed. M. Petschenig, CSEL 13. Vienna: Verlag der Österreichischen Akademie der Wissenschaften.

Causse, Antonin. 1947. De la Jérusalem terrestre à la Jérusalem céleste. *Revue d'Histoire et de Philosophie Réligieuses* 27: 12–36.

Cohn, Norman. 1993. *The Pursuit of the Millenium: Revolutionary Millenarians and Mystical Anarchists of the Middle Ages*, 3rd ed. London: Pimlico.

de Lubac, Henri. 1998–2009. *Medieval Exegesis: The Four Senses of Scripture*, 3 vols. Grand Rapids: Eerdmans.

Doren, Alfred. 1927. Wunschräume und Wunschzeiten. In *Vorträge der Bibliothek Warburg: 1924–1925*, ed. Fritz Saxl, 158–205. Leipzig: Teubner.

Elias, Norbert. 1982. Thomas Morus' Staatskritik. In *Utopieforschung: Interdisziplinäre Studien zur neuzeitlichen Utopie*, vol. 2, ed. Wilhelm Voßkamp, 101–150. Stuttgart: Metzler.

Frye, Northrop. 1965. Varieties of Literary Utopias. *Daedalus* 94 (2): 323–347.

Funke, Hans-Günter. 1991. Utopie, Utopiste. In *Handbuch politisch-sozialer Grundbegriffe in Frankreich: 1680–1820*, vol. 11, eds. Rolf Reichardt and Hans-Jürgen Lüsebrink, 6–100. München: Oldenbourg.

Graus, František. 1967. Social Utopias in the Middle Ages, trans. B. Standring. *Past & Present* 38: 3–19.

Grundmann, Herbert. 1927. *Studien über Joachim von Floris*. Leipzig: Teubner.

———. 1950. *Neue Forschungen über Joachim von Fiore*. Marburg: Simons.

Hartmann, Heiko. 2010. Utopias/Utopian Thought. In *Handbook of Medieval Studies: Terms—Methods—Trends*, vol. 2, ed. Albrecht Classen, 1400–1408. Berlin: De Gruyter.

Hartmann, Heiko, and Werner Röcke (eds.). 2013a. *Utopie im Mittelalter: Begriff – Formen – Funktionen*. Berlin: Akademie Verlag.

———. 2013b. Einleitung: Das Mittelalter – ein 'utopiegeschichtliches Vakuum'? In *Utopie im Mittelalter*, eds. H. Hartmann and W. Röcke, 3–27.

Isidore of Sevilla. 1911. *Etymologiae*, 2 vols., ed. Wallace M. Lindsay. Oxford: Clarendon Press.

James, Bruno Scott. 1952. *The Letters of St. Bernard of Clairvaux*. London: Bruns, Oates, and Washbourne.

Jaspert, Nikolas. 2001. Vergegenwärtigungen Jerusalems in Architektur und Reliquienkult. In *Jerusalem im Hoch- und Spätmittelalter. Konflikte und Konfliktbewältigung – Vorstellungen und Vergegenwärtigungen*, eds. Dieter R. Bauer, Klaus Herbers, and Nikolas Jaspert, 219–270. Frankfurt: Campus.

Joachim of Fiore. 1953. *Liber figurarum*, eds. L. Tondelli, M. Reeves, and B. Hirsch-Reich. Turin: Società Editrice Internazionale.

Kamlah, Wilhelm. 1969. *Utopie, Eschatologie, Geschichtsteleologie: Kritische Untersuchungen zum Ursprung und zum futurischen Denken der Neuzeit*. Mannheim: Bibliographisches Institut.

Konrad, Robert. 1965. Das himmlische und das irdische Jerusalem im mittelalterlichen Denken: Mystische Vorstellung und geschichtliche Wirkung. In *Speculum historiale: Geschichte im Spiegel von Geschichtsschreibung und Geschichtsdeutung*, eds. Clemens Bauer, Laetitia Boehm, and Max Müller, 523–540. Freiburg: Karl Alber.

Koselleck, Reinhard. 1982. Die Verzeitlichung der Utopie. In *Utopieforschung: Interdisziplinäre Studien zur neuzeitlichen Utopie*, vol. 3, ed. Wilhelm Voßkamp, 1–14. Stuttgart: Metzler.

Kühnel, Bianca. 1987. *From the Earthly to the Heavenly Jerusalem: Representations of the Holy City in Christian Art of the First Millenium*. Rom: Herder.

Landes, Richard A., Andrew C. Gow, and David C. van Meter (eds.). 2003. *The Apocalyptic Year 1000: Religious Expectation and Social Change*, 950–1050. Oxford: Oxford University Press.

Lochrie, Karma. 2016. *Nowhere in the Middle Ages*. Philadelphia: University of Pennsylvania Press.

Lovejoy, Arthur O., and George Boas. 1935. *Primitivism and Related Ideas in Antiquity*. Baltimore: John Hopkins Press.

Mannheim, Karl. 1954. *Ideology and Utopia: An Introduction to the Sociology of Knowledge*. New York: Harcourt, Brace.

McGinn, Bernard. 1979a. *Visions of the End: Apocalyptic Traditions in the Middle Ages*. New York: Columbia University Press.

———. 1979b. *Apocalyptic Spirituality. Treatises and Letters of Lactantius, Adso of Montier-en-Der, Joachim of Fiore, the Franciscan Spirituals, Savonarola*. Mahwa, NJ: Paulist Press.

———. 1985. *The Calabrian Abbot: Joachim of Fiore in the History of Western Thought*. New York: Macmillan.

———. 1992. Alter Moyses: The Role of Bernard of Clairvaux in the Thought of Joachim of Fiore. In *Bernardus Magister*, ed. John R. Sommerfeldt, 429–448. Aldershot: Variorum.

——— (ed.). 1998. *The Encyclopedia of Apocalypticism*, vol. 2, *Apocalypticism in Western History and Culture*. New York: Continuum.

———. 2001. The Apocalyptic Imagination in the Middle Ages. In *Ende und Vollendung: Eschatologische Perspektiven im Mittelalter*, eds. Jan A. Aertsen and Martin Pickavé, 79–94. Berlin: De Gruyter.

McNeill, David O. 1975. *Guillaume Budé and Humanism in the Reign of Francis I*. Geneva: Librairie Droz.

More, Thomas. 1995. *Utopia*, eds. George M. Logan, Robert M. Adams, and Clarence H. Miller. Cambridge: Cambridge University Press.

Müller, Götz. 1989. *Gegenwelten: Die Utopie in der deutschen Literatur*. Stuttgart: Metzler.

Neusüss, Arnhelm. 1972. Schwierigkeiten einer Soziologie des utopischen Denkens. In *Utopie: Begriff und Phänomen des Utopischen*, 2nd ed., ed. Arnhelm Neusüss, 13–122. Neuwied: Hermann Luchterhand.

Nipperdey, Thomas. 1975. Die Utopia des Thomas Morus und der Beginn der Neuzeit. In *Reformation, Revolution, Utopie: Studien zum 16. Jahrhundert*, 113–142. Göttingen: Vandenhoeck & Ruprecht.

Oexle, Otto Gerhard. 1977. Utopisches Denken im Mittelalter: Pierre Dubois. *Historische Zeitschrift* 224 (2): 293–339.

———. 1997. Utopie. *Lexikon des Mittelalters* 8: 1345–1348.

———. 1994: Wunschräume und Wunschzeiten: Entstehung und Funktion utopischen Denkens in Mittelalter, Früher Neuzeit und Moderne. In *Die Wahrheit des Nirgendwo: Zur Geschichte und Zukunft utopischen Denkens*, ed. Jörg Callies, 33–83. Rehburg-Loccum: Evang. Akad. Loccum, Protokollstelle.

Paviot, Jacques (ed.). 2014. *Les Projets de Croisade: Géostratégie et Diplomatie Européenne du XIVe au XVIIe Siècle*. Toulouse: Méridiennes.

Pranger, Marinus. 1994. *Bernard of Clairvaux and the Shape of Monastic Thought: Broken Dreams*. Leiden: Brill.

Raedts, Peter G.J.M. 1994. St. Bernard of Clairvaux and Jerusalem. In *Prophecy and Eschatology*, ed. Michael J. Wilks, 169–182. Oxford: Blackwell.

Reeves, Marjorie Reeves. 1976. *Joachim of Fiore and the Prophetic Future*. New York: Harper.

Reeves, Majorie, and Beatrix Hirsch-Reich. 1972. *The Figurae of Joachim of Fiore*. Oxford: Clarendon.

Renna, Thomas. 2002. *Jerusalem in Medieval Thought: 400–1300*. Lewiston, NY: Edwin Mellen Press.

Rexroth, Frank. 2008. Pierre Dubois und das Projekt einer universalen Heilig-Land-Stiftung. In *Gestiftete Zukunft im mittelalterlichen Europa*, eds. Frank Rexroth and Wolfgang Huschner, 309–331. Berlin: Akademie Verlag.

Riedl, Matthias. 2004. *Joachim von Fiore: Denker der vollendeten Menschheit*. Würzburg: Königshausen & Neumann.

———. 2012. A Collective Messiah: Joachim of Fiore's Constitution of Future Society. *Mirabilia: Revista Eletrónica de História Antiga e Medieval* 14: 57–80.

Saage, Richard. 2000. 'Christliche Utopie' – ein Widerspruch in sich selbst? Zum Verhältnis von Utopie und Chiliasmus. In *Zwischen Anfang und Ende: Nachdenken über Zeit, Hoffnung und Geschichte. Ein Symposium (Münster, Mai 1999)*, eds. Hermann Fechtrup, Friedbert Schulze, and Thomas Sternberg, 81–113. Münster: Lit.

———. 2001–2003. *Utopische Profile*, 4 vols. Münster: Lit.

———. 2013. Ist der Chiliasmus eine Utopie? Das Problem der Systemüberwindung in der Frühen Neuzeit bei Morus und Müntzer. In *Utopie im Mittelalter*, eds. H. Hartmann and W. Röcke, 167–182.

Schmitt, Jean-Claude. 2000. Appropriating the Future. In *Medieval Futures: Attitudes to the Future in the Middle Ages*, eds. John Anthony Burrow and Ian P. Wei, 3–18. Woodbridge: Boydell.

Schölderle, Thomas. 2011. *Utopia und Utopie: Thomas Morus, die Geschichte der Utopie und die Kontroverse um ihren Begriff*. Baden-Baden: NOMOS.

Seibt, Ferdinand. 1969. Utopie im Mittelalter. *Historische Zeitschrift* 208 (3): 555–594.

———. 1980. Liber Figurarum XII and the Classical Ideal of Utopia. In *Prophecy and Millenarianism: Essays in Honour of Marjorie Reeves*, ed. Ann Williams, 259–266. London: Longman.

———. 2001. *Utopica: Zukunftsvisionen aus der Vergangenheit*. München: Orbis.

———. 2002. *Die Begründung Europas: Ein Zwischenbericht über die letzten tausend Jahre*. Frankfurt am Main: S. Fischer.

Siebers, Johan. 2012. Noch-Nicht. In *Bloch-Wörterbuch: Leitbegriffe der Philosophie Ernst Blochs*, eds. Beat Dietschy, Doris Zeilinger, and Rainer E. Zimmermann, 403–412. Berlin: De Gruyter.

Stock, Markus. 2013. Paradoxer Gewinn: Raumpoetik und utopische Anschaulichkeit in Ulrichs von Etzenbach 'Alexander'-Anhang. In *Utopie im Mittelalter*, eds. H. Hartmann and W. Röcke, 113–128.

Stroumsa, Guy G. 1999. Mystische Jerusaleme. In *Kanon und Kultur: Zwei Studien zur Hermeneutik des antiken Christentums*, 31–66. Berlin: De Gruyter.

Thompson, Augustine. 1982. A Reinterpretation of Joachim of Fiore's *Dispositio novi Ordinis* from the *Liber Figuarum* [*sic*]. *Citeaux: Commentarii Cistercienses* 33: 195–205.

Tomasek, Tomas. 2001/2002. Zur Poetik des Utopischen im Hoch- und Spätmittelalter. *Jahrbuch der Oswald von Wolkenstein Gesellschaft* 13: 179–193.

Töpfer, Bernhard. 1964. *Das kommende Reich des Friedens: Zur Entwicklung chiliastischer Zukunftshoffnungen im Hochmittelalter*. Berlin: Akademie-Verlag.

Vieira, Fátima. 2010. The Concept of Utopia. In *The Cambridge Companion to Utopian Literature*, ed. Gregory Claeys, 3–27. Cambridge: Cambridge University Press.

Voigts, Michael C. 2014. *Letters of Ascent: Spiritual Direction in the Letters of Bernard of Clairvaux*. Cambridge: James Clarke & Co.

Vondung, Klaus. 2000. *The Apocalypse in Germany*. Columbia: University of Missouri Press.

Zudeick, Peter. 2012. Utopie. In *Bloch-Wörterbuch: Leitbegriffe der Philosophie Ernst Blochs*, eds. Beat Dietschy, Doris Zeilinger, and Rainer E. Zimmermann, 663–664. Berlin: De Gruyter.

CHAPTER 4

Living on the Edge of Time: Temporal Patterns and Irregularities in Byzantine Historical Apocalypses

András Kraft

The future is generally considered to be unknown. Yet, it is the presumably unknown future towards which Christian piety is ultimately directed. Christian spirituality is essentially teleological: it interprets actions and events from the perspective of the Last Judgment, which marks the end of the present world. "Eschatology is not just one particular section of the Christian theological system, but rather its basis and foundation, its guiding and inspiring principle. ... *Christianity is essentially eschatological...*" (Florovsky 1956: 27). Byzantium, being the Christian Roman Empire, enjoyed, therefore, an eschatological trajectory, which fostered at least as much anxiety as it engendered fascination.

The fascination with the end times was driven by a number of factors. For one, it was motivated by a universal curiosity about the future that calls for reliable foresight and proactive care. The prestige and authority of prophetical revelations further contributed to the appeal of the

A. Kraft (✉)
Medieval Studies Department, Central European University,
Budapest, Hungary

© The Author(s) 2017
S. Baumbach et al. (eds.), *The Fascination with Unknown Time*,
DOI 10.1007/978-3-319-66438-5_4

end times. Also, the verbally expressive if ambiguous language of word-images and symbols gave apocalypses an aesthetic and cognitive charm that inspired, among others, also the Byzantines. In addition, the often comforting tone and the use of familiar and thus dependable sequences of events bestowed trust upon the audience, whereas innovative twists and sudden changes in the narrative intensified their amazement.

Moreover, the Byzantine curiosity concerning the last times was invigorated by the fact that the Scriptures as well as official orthodox tradition are extremely reticent on this matter, especially for the period leading up to the Second Coming of Christ. To fill this gap, late antique and medieval authors produced a wide range of pseudonymous writings. These pseudepigrapha addressed the need to learn more about eschatological events. As a result, numerous prophetic texts were composed that addressed different sensitivities and different audiences: (1) *historical apocalypses* presented forecasts concerning the political fortunes of the empire, and (2) *moral apocalypses* disclosed information concerning the personal fate of the deceased in the afterlife. In this chapter, I focus on the former subgenre of Byzantine historical apocalypses, with the source material extending from the sixth to the thirteenth century, that is, from the early to the end of the middle Byzantine period. I do not discuss *moral apocalypses* here because a preliminary discussion on the perception of time in this literary subgenre already exists (Baun 2007: 144–147).

Historical apocalypses present narratives of the future up until the Last Judgment. Accordingly, I use the term *apocalyptic* in this genre-specific sense: an apocalypse reveals what is to take place in the last days of *this* world. Consequently, historical apocalypses can be considered a kind of historiography, and for this reason contemporary scholarship has 'mined' pseudepigraphal prophecies for new, otherwise unknown, historical information, following in the footsteps of the pioneering work by P. Alexander (1968). Furthermore, these prophecies have been examined as motivating causes behind political decision making (e.g., see Treadgold 2004; Magdalino 2007). Byzantine historical apocalypses have thus, first and foremost, been studied with a fact-oriented, historicist concern.

The present study proposes a different approach: it investigates how Byzantine apocalypses convey the concept of time to their audience. For this investigation, I have chosen a threefold approach: First, I sketch the chronology of the end times, which consists of various presumed dates for the Last Judgement and of a sequence of expected events that were anticipated to take place beforehand. Second, I discuss the typological structure of apocalyptic time, which underlies much of the prophetic accounts, as is shown by a few selected examples. Third, I reflect upon the speed of apocalyptic time by arguing for the significance of fluctuations in the narrative speed and by discussing the apocalyptic phenomenon known as the shortening of days.

When seen together, these three approaches help reconstruct a rather detailed picture of the content and structure of the 'not so unknown' future. In fact, the Byzantines had a fairly clear idea of how their future history would play out and how it would come to an end. Arguably, this standard future narrative conditioned their confidence to live at the very edge of time (cf. Alexander 1962: 344–345).

THE CHRONOLOGY OF APOCALYPTIC TIME

The apocalyptic tradition in Byzantium provided its audience with a more or less coherent chronology of the 'not so unknown' future. Although it left much room for ambiguity and interpretation, it did present a standardised vision of the future that was composed of demonstrative calculations at one end and of revelatory predictions at the other.

Calculations of the Date of the Apocalypse

Despite the biblical assertion (Acts 1:7; cf. Mt 24:36, Mk 13:32, 1 Thess 5:1–2, Rv 16:15) that the precise time of the end is unknowable, Byzantine apocalyptists continuously professed to reveal the exact year of the Last Judgment. They supported their claims with various calculations. The most popular computational scheme was based on the presumption that the age of the world would not exceed 6000 (alternatively, 6500 or 7000) years after the creation, which, according to a reckoning widely held by the tenth century, had taken place on 1 September 5509 BCE (Grumel 1958: 111–128). The assumption that the world will last only 6000 years was based on a synoptic reading of Ps 90:4 (KJV): "For a thousand years in thy sight are but as yesterday when it is past, and as a watch in the night" (cf. 2 Pet 3:8) and Gen 1:1–2:3, wherein creation is said to have lasted six days (see Podskalsky 1974: 356n23). Thus, the reasoning went: if God created the world in six days, and if six days for God amount to 6000 years to mankind, then this world will come to an end after 6000 years have lapsed, which early Christians expected to take place around the year 500 CE (Brandes 1997). For instance, in a Greek redaction of the *Tiburtine Sibyl*, named by its modern editor P. Alexander the *Oracle of Baalbek*, it is predicted that Constantinople will fall around the year 510 CE: "Do not boast, city of Byzantium, thou shalt not hold imperial sway for thrice sixty of thy years" (translated by Alexander 1967: 25). If one adds thrice sixty years (i.e., 180 years) to the year 330 CE (the year Constantinople was consecrated), one arrives at the date 510 CE.

Apocalyptic expectations indeed peaked around the year 500 CE (Brandes 1997: 53–63) as well as around the year 1000 CE. Niketas David the Paphlagonian, for instance, advanced various calculations, of which one demonstrates that the end of the world would arrive around the significant number 6500 ($= 991/992$ CE), which marked the mid-point of the seventh millennium according to the Byzantine calendar (ed. Westerink 1975: 192, ll.30–38; Mango 1984: 435–436; cf. Magdalino 2003: 269; Brandes 2011: 314). I. Ševčenko (2002) and P. Magdalino (2002) have published and discussed further evidence that explicitly promoted the view that the end of the world would occur a thousand years after Christ's incarnation (or resurrection). There is also the famous testimony by John Tzetzes about an oracle that predicted how Constantinople would be destroyed before its millennial anniversary: "Woe you, o Seven-Hilled City, for thou shall not be a thousand years old" (*Ioannis Tzetzae Historiae* 1.663, ed. Leone 1968: 370 [my translation]; *Ioannis Tzetzae Epistulae* 1.9, ed. Leone 1972: 88; Mango 1980: 212). The destruction of the imperial capital would be closely followed by the end of the world as evinced by a number of Byzantine apocalypses (e.g., *Last Vision of the Prophet Daniel* §69–86, ed. Schmoldt 1972: 138–144). Finally, following the Ottoman conquest of Constantinople in 1453, apocalyptic speculations climaxed with even the Patriarch, Gennadius Scholarius, endorsing the view that the apocalypse was to occur shortly, namely at the end of the seventh millennium, that is, around the year 1492 CE (Petit et al. 1935: 511–512; Turner 1964: 369–371; Podskalsky 1974: 357).

These computational predictions could, at best, propose a likely date, which, however, repeatedly failed to materialise. The New Testament advisory thus continued to remain valid insofar as the date of the end could not be known in advance. A more reliable means to learn about the unknown time of the apocalypse was to study the signs of the end and their sequence.

The Sequence of Future Events

Apocalyptic anxieties were often triggered and reinforced by natural and moral signs of the end. Natural signs included catastrophes such as earthquakes and famines (Mt 24:7, Lk 21:11), whereas moral signs revolved around pseudo-prophets who would pervert orthodoxy (Mt 24:11, Mk 13:22, Rv 19:20; see further Brandes 1997: 45). These various signs were structured into a successive account, which by the eighth century had developed into a standard narrative of the future that was continuously reused and updated—with central themes persisting even until today. This standard narrative is best put forward in prophecies that are generally attributed to the Prophet Daniel. These pseudo-Danielic prophecies (for their respective bibliographies, see DiTommaso 2005: 347–374) presented a popular and widespread literary sub-genre in which apocalyptists disclosed the future succession of political events. After an initial period of hardship, usually suffered at the hands of a Muslim foe, the appearance of a series of Byzantine emperors is described, who were to carry out a particular set of eschatological tasks, namely (1) to defeat all foreign enemies, in particular the Muslim foe, with or without the help of the Blond Nations, which referred, at times, to the Russians, at other times, to the Latins (occasionally, an emperor would also have to defend or reconquer the city of Constantinople); (2) to inaugurate a subsequent period of peace and prosperity, with benefactions granted to the people and the Church; and (3) to travel to Jerusalem, where the last of these emperors would abdicate his imperial dignity to Christ. It depended on each apocalyptist's discretion to reveal how many emperors would accomplish these tasks. Adjacent to the emperor's abdication, either preceding or succeeding it, was the arrival and defeat of the eschatological peoples of the north, which evolved from a biblical motif (Ez 38–39) into an apocalyptic *topos*, starting in the fourth century (van Donzel and Schmidt 2009: 16–31). All these events were embellished with the potent imagery of the calamities from the book of Revelation and the Synoptic apocalypse (Mt 24, Mk 13, Lk 21).

The final role in the end-time narrative was reserved for the Antichrist, whose origin, deeds, and eventual destruction would be described in varying detail. As a rule of thumb it can be said that the later the prophecy, the shorter the descriptions of the Antichrist. In late Byzantine pseudo-Danielic prophecies the attention rests almost exclusively with the fortunes of the empire and the emperor(s). It appears that with the increasing decline of imperial power, the curiosities and anxieties concerning the arrival of the Antichrist were steadily alleviated. One needs to remember that the Antichrist was generally believed to be a future emperor, as attested by the *Commentary on Revelation* by

Andrew of Caesarea (d. 614) (ed. Schmid 1955: 136–137, 189) and by a number of prophecies such as the *Apocalypse of Leo of Constantinople* (ed. Maisano 1975: 90, ll.430–435, 551–552). That is to say, the decrease of imperial power might have minimised the unease and ambiguity concerning the emperor as being both eventual saviour and potential Son of Perdition. This reduction of anxiety stands in stark contrast with developments in the West, where the same ambiguity—associated with the papacy—intensified throughout the High Middle Ages with the increase of papal authority (McGinn 1978).

With the defeat of the Antichrist, the standard narrative of the future comes to an end. Historical apocalypses seldom go beyond this point (a notable exception being the *Apocalypse of Leo of Constantinople*, ed. Maisano 1975: 100–112). Based on this sequence of events, a perceptive Byzantine would have felt competent to discern how close was the end of time. For instance, if there had not yet been an imperial abdication in Jerusalem, then the end was, arguably, not that imminent.

THE TYPOLOGICAL STRUCTURE OF APOCALYPTIC TIME

When reading Byzantine historical apocalypses, one is presented with an orderly flow of past and future events. At first sight, the various events appear similar to dresses on a clothesline (to use a metaphor), which has a definitive start and end; the start is either the creation of the world or any subsequent event and the end is the Second Coming. These dresses or events, however, are not only structured along the linear thread of chronology, they are also—and arguably more importantly—structured along the overlying pattern of typology. With regard to our metaphor this would mean that the dresses are hung up at more than one place on the clothesline.

Typology is a mode of thought (Goppelt 1964: 332; Frye 1982: 80) that was used, first and foremost, in biblical hermeneutics to uncover correspondences between Old Testament (OT) and New Testament (NT) characters and events. Typology relates an OT historical adumbration, the *type*, with its NT fulfilment, the *antitype*. Such correspondence was understood to reflect divine providence, which directs and structures the course of history in such a way that its final fulfilment is repeatedly prefigured in earlier announcements. The focal point of all NT typologies is Christ, in whom the final consummation of the world was expected to take place. Accordingly, Christ has been presented and explained as a new Adam (Rom 5:14), as a new David (Mt 1:1–17, Acts 2:29–32), as a new

Jonah (Mt 12:39–42, Lk 11:29–32), and the like. An OT type develops into its subsequent fulfilment, its NT antitype, which always presents an intensification of the original type and, thus, carries an eschatological meaning (Goppelt 1964: 330, 334–344; Davidson 2011: 11). This typological reasoning does not stop with the last book of the Bible or with patristic exegesis. When writing the 'history of the future,' Byzantine apocalyptists continued to reveal typologies that connect future (as well as present) events or characters with Old *and* New Testament types. In the following, I illustrate this point with a few notable examples.

Massacre of the Innocents Typology

The *Diegesis Danielis*, a prophecy composed sometime in the eighth or ninth century (Berger 1976: 36; Zervos 1983: 756–757; DiTommaso 2005: 130–141; cf. Aerts 2010: 468), begins with an oracle, which probably relates events surrounding the second Arab siege of Constantinople in 717/718. In this prophecy, three Arab armies are said to approach the imperial capital on different routes. The following is said about one of these armies:

Diegesis Danielis II.5–8, ed. Berger 1976: 12	Mt 2:16
(5) καὶ ὁ ἕτερος (scil. υἱὸς τῆς Ἄγαρ) ἔλθη ἐπὶ τὸ μέρος τῆς Περσίδος καὶ τὴν χώραν τὴν Γαλιλαίαν, Ἀρμενίας τὸ ἄκρον καὶ πόλιν Τραπεζοῦντα. (6) καὶ ἔλθη ἐπὶ τὸ μέρος τῆς γῆς τῶν Μερόπων. (7) καὶ κατακόψει ἄρρενας παῖδας ἀπὸ διέτους καὶ τριέτους ἄνωθεν. (8) καὶ ἀναλώσει ἐν μαχαίρᾳ πλήθη πολλή. [my emphasis]	τότε Ἡρῴδης ἰδὼν ὅτι ἐνεπαίχθη ὑπὸ τῶν μάγων ἐθυμώθη λίαν, καὶ ἀποστείλας ἀνεῖλεν *πάντας τοὺς παῖδας τοὺς ἐν Βηθλέεμ καὶ ἐν πᾶσι τοῖς ὁρίοις αὐτῆς ἀπὸ διετοῦς καὶ κατωτέρω, κατὰ τὸν χρόνον ὃν ἠκρίβωσεν παρὰ τῶν μάγων.* [my emphasis]
(5) and the second (i.e., son of Hagar) will attack the region of Persia and the land of Galilee, the mountain top of Armenia, and the city of Trebizond. And he will attack the region of the land of the Meropes[1] (i.e., the island of Kos) *and will cut down male children from two and three years of age and above.* And he will kill by sword a great multitude. [my translation and emphasis]	Then Herod, when he saw that he was mocked of the wise men, was exceeding wroth, and sent forth, *and slew all the children that were in Bethlehem, and in all the coasts thereof, from two years old and under,* according to the time which he had diligently inquired of the wise men. (KJV [my emphasis])

[1]Berger understands the "land of the Meropes" to designate the island of Kos (1976: 51–52); see also Halliday (1913: 279). Alternatively, it could be an unprecedentedly early reference to—or a later interpolation of—Thrace, which is otherwise only attested in the fourteenth century, see Soustal (1991: 51, 354).

The description of the Arab advance appears to evoke the massacre of the innocents as known from Mt 2:16. The attentive reader will have noticed a slight change: the NT type recounts that all male children *under* (κατωτέρω, i.e., "younger than") two years were killed, whereas the Byzantine apocalypse foretells that all male children *above* (ἄνωθεν, i.e., "older than") that age are killed (in Byzantine Greek ἄνωθεν means "above," as correctly pointed out by Aerts 2010: 469; and *pace* Zervos 1983: 763). It appears that the Byzantine apocalyptist inverted here one element of the historical correspondence und turned it into its antithetical or anti-symmetrical counterpart.

Exodus Typology

The same kind of inversion can be observed at the beginning of the *Last Vision of the Prophet Daniel*, which was probably composed in response to the Fall of Constantinople to the Latin Crusaders in 1204 (Brandes 2007: 253). This prophecy starts as follows:

Last Vision of the Prophet Daniel §1, 11–13, ed. Schmoldt 1972: 122, 126	Ex 10:3–5
(1) τάδε λέγει κύριος παντοκράτωρ· οὐαί σοι γῆ … (11) καὶ οὐαί σοι γῆ ἐκ τῶν βασάνων ὧν μέλλει ἐξαποστεῖλαι κύριος παντοκράτωρ ἐπί σε. (12) *ἀκρίδας ἀγρίας* καὶ *ἀναιμάκτους* μέλλει πέμψαι ἐπί σε. (13) καὶ οὔτε ζῶον οὔτε δένδρον μέλλουσιν ἄψασθαι εἰ μὴ τοὺς μὴ μετανοήσαντας διὰ τὰς πολλὰς αὐτῶν ἀνομίας καὶ ἀδικίας. [my emphasis]	(3) εἰσῆλθεν δὲ Μωυσῆς καὶ Ααρων ἐναντίον Φαραω καὶ εἶπαν αὐτῷ *Τάδε λέγει κύριος ὁ θεὸς τῶν Εβραίων* Ἕως τίνος οὐ βούλει ἐντραπῆναί με; ἐξαπόστειλον τὸν λαόν μου, ἵνα λατρεύσωσίν μοι. (4) ἐὰν δὲ μὴ θέλῃς σὺ ἐξαποστεῖλαι τὸν λαόν μου, ἰδοὺ ἐγὼ ἐπάγω ταύτην τὴν ὥραν αὔριον *ἀκρίδα* πολλὴν ἐπὶ πάντα τὰ ὅριά σου, (5) καὶ καλύψει τὴν ὄψιν τῆς γῆς, καὶ οὐ δυνήσῃ κατιδεῖν τὴν γῆν, καὶ κατέδεται πᾶν τὸ περισσὸν τῆς γῆς τὸ καταλειφθέν, ὃ κατέλιπεν ὑμῖν ἡ χάλαζα, *καὶ κατέδεται πᾶν ξύλον τὸ φυόμενον ὑμῖν ἐπὶ τῆς γῆς.* [my emphasis]

(Continued)

(1) *This saith the Lord Almighty:* Woe you earth ... (11) And woe you earth because of the trials which the Lord Almighty will send upon you. (12) He will send upon you wild and bloodless *locusts.* (13) *And they will not touch either animal or tree* except for those who did not repent for their great lawlessness and injustice. [my translation and emphasis]

(3) And Moses and Aaron came in unto Pharaoh, and said unto him, *Thus saith the Lord God of the Hebrews,* How long wilt thou refuse to humble thyself before me? Let my people go, that they may serve me. (4) Else, if thou refuse to let my people go, behold, *tomorrow will I bring the locusts into thy coast:* (5) And they shall cover the face of the earth, that one cannot be able to see the earth: and they shall eat the residue of that which is escaped, which remaineth unto you from the hail, *and shall eat every tree which groweth for you out of the field.* (KJV [my emphasis])

This pseudo-Danielic apocalypse reverberates prophetic material from the book of Revelation, notably the image of angels bringing calamities (Rv 8–11), while also adopting a typology from Rv 9:3–4 that contains an anti-symmetrical correspondence with the book of Exodus (cf. Goppelt 1966: 238). The typology revolves around God sending a number of afflictions upon the Earth through the mediation of Moses (in the case of Exodus) or through angels (in Revelation and in the *Last Vision of the Prophet Daniel*). One example of these afflictions is the sending of locusts, which devour everything (in the case of Exodus) or *not* everything but only the sinful (in Revelation and in the *Last Vision of the Prophet Daniel*). That is to say, the Exodus type is not fully congruent with the apocalyptic antitype. The latter is constructed in close resemblance but in a somewhat inverted fashion: in the end times, the locusts will not devour everything indiscriminately as they did during the age of Moses. This inversion merely qualifies the typological correspondence that relates the oppression of God's chosen people at the end of times with events experienced during the Egyptian captivity.

In all likelihood, this particular typology was deliberately taken over from Revelation. Internal evidence suggests that the *Last Vision of the Prophet Daniel* was composed in the middle of the thirteenth century during the Latin occupation of Constantinople while the Βασιλεία τῶν Ῥωμαίων was, in fact, in captivity in the Byzantine successor states in Western Greece and in Asia Minor. It is probable that this typology voices a more widespread belief of the thirteenth century that saw the Byzantines as experiencing the eschatological equivalence of the

Egyptian captivity. Collaborative evidence for this view can be found in a panegyric by the historian Niketas Choniates (d. 1217), who eulogised the Byzantine emperor in exile, Theodore I. Laskaris (d. 1221), as a new Moses and new Zorobabel (*Nicetae Choniatae Orationes et Epistulae* ll.1–7, ed. Van Dieten 1972: 147; see further Angelov 2007: 86, 99).

Flood Typology

Furthermore, the apocalyptic section in the *Life of Andrew the Fool* (ed. and trans. Rydén 1995, vol. 2: 258–285; ed. and trans. ibid. 1974), which is a fictitious hagiographic work from the mid-tenth century, presents a number of examples for apocalyptic typologies. For instance, it reverberates and expands the NT flood typology from Mt 24:37–38 and Lk 17:26–27, where the eve of the end times is likened to the peace and quiet of the period preceding the Great Flood (Gn 6–9). Building upon this typology, the apocalyptist visionary foretells the following:

Apocalypse of Andrew the Fool 853 B–C, 856C, ed. Rydén 1974: 202–203 (trans. ibid. 215–217)

(853B) Ἐν ταῖς ἐσχάταις ἡμέραις ἀναστήσει κύριος ὁ θεὸς βασιλέα ἀπὸ πενίας καὶ πορεύσεται ἐν δικαιοσύνη πολλῇ καὶ πάντα πόλεμον παύσει καὶ τοὺς πένητας πλουτίσει καὶ ἔσται ὡς ἐπὶ τοῦ Νῶε τὰ ἔτη. ... ἔσονται γὰρ οἱ ἄνθρωποι κατὰ τὰς ἡμέρας αὐτοῦ πλούσιοι σφόδρα καὶ ἐν εἰρήνη ἀπείρῳ 'τρώγοντες καὶ πίνοντες, γαμοῦντες καὶ ἐκγαμίζοντες' καὶ ἐν ἀφοβίᾳ πολέμου ... (856C) Καὶ ἔσται πολλὴ χαρὰ τότε καὶ ἀγαλλίασις, καὶ ἀγαθὰ ἀπὸ τῆς γῆς καὶ ἀπὸ τῆς θαλάσσης ἀνατελεῖ πλούσια. Καὶ ἔσται ὃν τρόπον ἦσαν ἐπὶ τοῦ Νῶε ἐν ἠρεμίᾳ εὐφραινόμενοι μέχρις οὗ ἦλθεν ὁ κατακλυσμός. [my emphasis]

In the last days the Lord God will raise up an emperor from poverty. He will walk in great righteousness and bring every war to an end and make the poor rich and *the years will be as in the time of Noah.* ... For in his days men will be very rich and in deep peace they will be 'eating and drinking, marrying and giving in marriage' without fear of war ... (856C) There will be great joy then and gladness. Good things will come up from the earth, and from the sea riches will rise. It will be as when *in the days of Noah* men enjoyed themselves in peace until the flood came. [my emphasis]

A blissful period in the, presumably, near future will see the repetition of the joys known from the antediluvian age. The reader is thus transferred from his or her contemporary world of the middle Byzantine period back to the age of the Patriarchs. This notion that an ideal emperor would usher in a time of peace and prosperity that is reminiscent of Noah's age derives from the seventh-century *Apocalypse of Pseudo-Methodius*, Chap. xiii.17, wherein this NT typology was first applied to a future Byzantine emperor (for the Syriac text, see *Die Syrische Apokalypse des Pseudo-Methodius*, ed. Reinink 1993: 41; and for the subsequent Greek translation, see *Die Apokalypse des Pseudo-Methodius*, ed. Aerts and Kortekaas (1998: 180).

Christological Typologies

Typological constructions pertaining to the anticipated future do not only relate to events but also to personal actors. As already indicated, NT typologies are most often Christocentric; they centre on indicating correspondences between Christ and OT precursors. Apocalyptic prophecies faithfully continue this Christocentric focus. In fact, the most apparent and pervasive typologies in the apocalyptic genre revolve around the literary figure of the Last Roman Emperor and the Antichrist, which are both constructed in explicit correspondence to Christ. The Anti-Christ is always presented as His antithetical or anti-symmetrical antitype, insofar as the Antichrist's deeds are an inversion of Christ's genuine miracle workings and teachings. In contrast, the Last Emperor is portrayed in close congruence with Christological characteristics that include, among others, righteousness, piety, humility, and associations with the Resurrection. The Last Emperor was, thus, conceptualised as a positive or symmetrical antitype of Christ. A most succinct expression of this underlying typological reference system can be found in a little known prophecy that was probably written in the early ninth century.

Anonymi de rebus Byzantinis vaticinium, ed. Vasiliev 1893: 48	Heb 7:3
… καὶ ἐξαναστήσεται αἰφνίδιος (*lege* αἰφνίδιον) βασιλεὺς δίκαιος ἀφωμοιωμένος τῷ υἱῷ τοῦ θεοῦ, … [my emphasis]	ἀπάτωρ ἀμήτωρ ἀγενεαλόγητος, μήτε ἀρχὴν ἡμερῶν μήτε ζωῆς τέλος ἔχων, ἀφωμοιωμένος δὲ τῷ υἱῷ τοῦ θεοῦ, μένει ἱερεὺς εἰς τὸ διηνεκές. [my emphasis]
… and suddenly a righteous king will rise up who *resembles the Son of God* … [my translation and emphasis]	Without father, without mother, without descent, having neither beginning of days, nor end of life; but made like unto the Son of God; abideth a priest continually. (KJV)

The way in which the ideal emperor is likened here to the Son of God reminds one of Heb 7:3, where—in the same words—the OT priest Melchizedek is likened to Christ (cf. Goppelt 1966: 196–205). The typological correspondance is, thus, continued into the future. One is presented with the following scheme:

Melchizedek	Christ	Last Emperor	Antichrist
OT type	NT antitype and NT type	Symmetrical antitype	Anti-symmetrical antitype

The figures of the Antichrist and the Last Emperor are constructed— persistently throughout the Byzantine apocalyptic tradition—in anti-symmetrical and symmetrical correspondence with Christ. As I have discussed this typological scheme elsewhere (Kraft, forthcoming), I refrain from going into more detail here.

These examples should suffice to demonstrate that numerous pro-tagonists and events in Byzantine apocalypses are typologically informed motifs. When weaving the carpet of providential history, the Byzantine apocalyptists used typological (be these symmetrical or anti-symmetrical) constructs that reverberated past events of salvation history (particu-larly Christological events) and sewed them into the fabric of canoni-cal and apocryphal material that pertains to the end times. That is, they drew on the past to interpret the future (see Schmidt-Biggemann 2004: 334–335).

The typological links were the threads onto which these various ele-ments of the apocalyptic repertoire could be sewn. If a Byzantine apoca-lyptic narrative appears as a linear clothesline (to continue the metaphor used earlier), then the clothesline represents chronology, the clothes-pins represent historical moments, and the dresses represent events, which can be fastened with more than one clothespin. Just as dresses on a clothesline, particular Christological events can appear more than once on the string of time. These typologies introduce nonlinear inter-weavings into the fabric of linear, successive history. As a result, there is no need for genuine novelties, nor for new ideologies or technological progress in the series of future events; there will be nothing fundamen-tally new in the future. What will occur are typological emulations that intensify and fulfil earlier prefigurations. This persuasion must have, in turn, provided confidence to the apocalyptic publicists in their continu-ous (re)writing of the history of the future, as the future was not alto-gether unknown. Yet, the relative familiarity with the anticipated events of the future did not lead to an invariable narrative, quite to the contrary.

New typologies, with their positive and antithetical characteristics, were continuously worked out, leading to ample variations and dramatising comparisons that must have sustained the unfading fascination with the reading and hearing about the end of times, as evidenced by the copious manuscript tradition.

THE SPEED OF APOCALYPTIC TIME

At a first glance, Byzantine historical prophecies present time as proceeding in regular chronological order. Yet, upon a more careful inspection, the continuous flow of time is marked by phenomenological and physical distortions. For one, the perception of apocalyptic time that these prophecies advance is dependent upon the narrative speed with which the events and personages are presented. Furthermore, the end time was expected to undergo physical distortions, with the natural duration of days and hours being suspended. In the following I elaborate on these two aspects that pertain to the speed of apocalyptic time.

Narrative Speed

As already mentioned, Byzantine apocalypses are historiographical narratives. Narratives describe events and characters at different lengths and therefore give the reader a different sense of magnitude with regard to significance as well as to the passing of time. In general, it can be said that the greater detail an author provides, the slower the narrative proceeds, and conversely, the lesser detail is provided, the quicker events appear to pass. More precisely, the perception of time that a narrative evokes depends on the relationship of the (1) narrated time—the duration of an event within the narrative—to the (2) narrating time—the duration it takes the narrator to recount the event. The ratio of these two variables establishes the narrative speed, which accelerates when events are presented in summary and decelerates when events are depicted scenically (Genette 1980: 93–95; Prince 1982: 54–59). Direct speech, for instance, is a scenic device that approximates the actual narrating time with the duration of the event and thus temporarily slows down the narrative speed (De Jong and Nünlist 2007: 11). Although many Byzantine apocalypses present the reader or listener with direct speeches such as divine commands and prayers, the general mode of presentation consists of summaries.

That being said, even within these summaries the narrative speed in Byzantine apocalypses can differ significantly. For instance, in the apocalyptic section of the *Life of Andrew the Fool* the narrator prophecies the deeds of a series of eschatological rulers, for whom we are given the anticipated lengths of their reigns. Each reign is described in varying detail. If we quantify the narrating time of each description with its word count and divide it by the narrated time, then we can establish a ratio that indicates the respective narrative speed of each description. The lower the ratio, the higher the speed. By comparing the different narrative speeds, one can establish the rhythm of the prophecy.

Apocalypse of Andrew the Fool, ed. Rydén 1995, vol. 2: 260–269				
Reference	Theme	Narrated time	Word count	Ratio
ll.3824–3858	A victorious emperor rules	32 years	365	11.4
ll.3859–3884	A son of lawlessness rules	3.5 years	270	77.1
ll.3885–3906	A pagan emperor rules	Unspecified	219	N/A
ll.3907–3912	An Ethiopian emperor rules	12 years	61	5.1
ll.3913–3920	An Arab emperor rules	1 year	82	82
ll.3921–3923	Three young men reign in peace	150 days	20	48.8

This prophecy shows a markedly irregular narrative speed that literally jumps between acceleration and deceleration. This irregularity is further amplified by the fact that within each narrated timeframe particular brief events are developed in greater detail. It remains an open question which effect these rapid shifts may have had on its audience. At this point, one might doubt whether these shifts are of any importance at all, because the narrated time periods used here comprise apocalyptically connoted numbers: 32, 12, 3.5; they are *topoi*. I would respond that the validity, historicity, and phenomenological impact of a motif (such as a time period) are not a priori diminished by its topical nature. *Topoi* are rhetorical devices that contextualise events and characters, which *can* be fictional; however, they are usually applied to historical facts (Pratsch 2005: 364–371). Moreover, the phenomenological impact of influencing the meta-textual sensation of a narration is not affected by issues of historicity. That is to say, these topical durations can be taken literally as an unbiased Byzantine audience would have done.

The fluctuating narrative speeds result in an erratic rhythm that must have evoked a curious uncertainty about what follows next. Such a rhythm might have served the aesthetic means of enforcing a fascinating anxiety with the future, as volatile shifts disorient an audience and, concomitantly, generate a climate of uncertainty that begs for clarification. Arguably, the urge to overcome this disorientation and ambivalence inflated the fascination with apocalyptic prophecies. Thus, the narrative speed of Byzantine apocalypses could phenomenologically influence one's perception of, and interest in, future time. It remains to be investigated if this thesis can be supported with other examples, which are, however, relatively rare. In contrast to the analysed section of the *Apocalypse of Andrew the Fool*, most apocalypses remain vague or even uncommitted on providing exact durations and thus frustrate such analysis.

The Shortening of Days

Even in the vaguest descriptions of temporal durations, the narrator presents the course of future history as a succession of periods that are composed of years of equal length. The duration of future time is understood to be just as uniform as the durations of present time. There is, however, one exceptional occurrence that upsets this uniformity: the phenomenon of the 'shortening of days,' which is based on the following Gospel account:

Mt 24:22

καὶ εἰ μὴ ἐκολοβώθησαν αἱ ἡμέραι ἐκεῖναι, οὐκ ἂν ἐσώθη πᾶσα σάρξ· διὰ δὲ τοὺς ἐκλεκτοὺς κολοβωθήσονται αἱ ἡμέραι ἐκεῖναι. (cf. Mk 13:20)

And except those days should be shortened, there should no flesh be saved: but for the elect's sake those days shall be shortened. (KJV)

This phenomenon is referred to in a number of Byzantine apocalypses as, for instance, in the early sixth-century *Oracle of Baalbek*, in the mid-thirteenth-century *Last Vision of the Prophet Daniel*, or in the *Apocalypse of Leo of Constantinople*, whose current form was probably redacted in the course of the twelfth century (ed. Maisano 1975: 20).

Oracle of Baalbek ll.178–180, ed. Alexander 1967: 20 (trans. ibid. 28)

ἐν δὲ τῇ ἐνάτῃ γενεᾷ κολοβωθησόνται τὰ ἔτη ὡσεὶ μῆνες καὶ οἱ μῆνες ὡσεὶ ἑβδομάδες καὶ ἑβδομάδες ὡς ἡμέραι καὶ ἡμέραι ὡσεὶῶρ αι.

In the ninth generation the years will be shortened like months, and the months like weeks, and the weeks like days, and the days like hours.

Last Vision of the Prophet Daniel §78–80, ed. Schmoldt 1972: 142

(78) καὶ κρατήσει ὁ τρισκατάρατος δαίμων ἔτη τρία ἥμισυ. (79) τότε ὁ χρόνος ὡς μὴν διαβήσεται, ὁ μὴν ὡς ἑβδομάς, ἡ ἑβδομὰς ὡς ἡμέρα, ἡ ἡμέρα ὡς ὥρα καὶ ἡ ὥρα ὡς στιγμὴ διὰ τοὺς ἐκλεκτοὺς καὶ δούλους τοῦ θεοῦ. (80) μετὰ δὲ τὴν συμπλήρωσιν τῶν τριῶν ἥμισυ χρόνων βρέξει ὁ θεὸς πῦρ ἐπὶ τὴν γῆν, καὶ κατακαήσεται ἡ γῆ πήχεις τρ ιάκο ν τα.

(78) And the thrice-accursed demon will rule for three and a half years. (79) Then the year will pass like a month, and the month like a week, the week like a day, and the day like an hour, and the hour like a moment for the sake of the elect and the servants of God. (80) After the completion of the three and a half years God will rain fire onto the earth and the earth will burn thirty cubits deep. [my translation]

Apocalypse of Leo of Constantinople ll.562–567, ed. Maisano, 1975: 98

ὁ δὲ φιλάνθρωπος Θεὸς κολοβώσειν ἔχει τὰ ἔτη καὶ τὰς ἡμέρας αὐτοῦ καὶ ποιήσει αὐτὰς ὀλίγας καὶ κωφάς, ὡς καὶ προεῖπον ἐγὼ Δανιὴλ καὶ προέθηκα, ὅτι ποιήσειν ἔχει τὰ ἔτη τρία ὡς μῆνας τρεῖς, καὶ τοὺς τρεῖς μῆνας ὡς τρεῖς ἑβδομάδας, καὶ τὰς τρεῖς ἑβδομάδας ὡς τρεῖς ἡμέρας, καὶ τὰς τρεῖς ἡμέρας ὡς ὥρας τρ εῖς.

But the benevolent God will shorten his [i.e., the Antichrist's] years and days and will make them few and empty, as I, Daniel, foretold and added that He will make three years as three months, and three months as three weeks, and three weeks as three days, and three days as three hours. [my translation]

The *Apocalypse of Leo of Constantinople* reaffirms the reason for the shortening of days that is given in Mt 24:22 (Mk 13:20), namely, that the days are shortened for the benefit of mankind. This is the predominant explanation provided in the Byzantine tradition, which might come as a surprise because a cosmological explanation would appear easy to arrive at, given the imagery of the "rolling up of the heavens" in Rv 6:14. With the collapse of the heavens, time would have to change, because time was generally considered—following Aristotle—to be measured movement (Aristotle, *Physics* IV.11, 219b1f). In particular, Plato had proposed that time be considered as dependent upon the movement of the celestial spheres (Plato, *Timaeus* 37d). Thus, if the heavens were to disintegrate, time would have to change concomitantly. The Byzantine commentaries on Revelation, however, avoid any such explanation. Most notably, there is no connection being made between the "rolling up of the heavens" and the "shortening of the days" in the commentary by Andrew of Caesarea (d. 614), which was the most authoritative exegesis on Revelation in Byzantium. Andrew understands the "rolling up" in four possible ways: it can refer to (1) the unknown time of the Second Coming; (2) the convulsive anguish suffered by the heavenly powers on observing human sinfulness (which is how Oecumenius

had understood this verse a few decades earlier; see Suggit 2006: 75);
(3) a change for the better; or (4) the complete revelation of the future
blessings (ed. Schmid 1955: 70–71; trans. Constantinou 2011: 99).
There was, therefore, not even an attempt to explain the shortening of
days in cosmological terms.

The apocalyptic tradition upholds that the days will be shortened
only for the benefit of the elect, so that suffering shall be minimised.
The underlying assumption was that suffering causes the perception
of time to slow down and, to ease this suffering, time would speed up.
There is ample evidence that the Byzantines experienced time to slow
down in times of hardship. A good witness is John Kaminiates who, in
his account on the sack of Thessaloniki in 904, talks about τοῦ καιροῦ
τὴν ἐπίτασιν, the stretching or prolongation of time (*De expugnatione
Thessalonicae* 1.91, ed. Böhlig 1973: 49), which he came to experience
amidst the horrors of carnage, exposure to the elements, thirst, and fear.
Conversely, the shortening—and thus the speeding up—of time would
diminish the suffering and was, therefore, understood as a philanthropic
act of relief.

CONCLUSION

The deliberate distortion of time is a temporal anomaly that is given
meaning in the ethical, and not in the physical, context of the end times.
Linear time can be compressed and altered. God can play with time.
This temporal irregularity is built into the ordered sequence of events
that otherwise knows only homogeneously measured and consecutively
structured intervals of time (32 years, 12 years, etc.). However, this
homogeneity is qualified by the author's narrative speed, which is used
to influence the audience's perception of the passage of time during particular events. Arguably, it concomitantly enforced the audience's fascination with the erratic and intermittent future.

Although canonical Scriptures and Church dogma were deeply
reserved on future matters, the Byzantines possessed an extracanonical
tradition that supported the attempt to calculate the date of the end.
Moreover, this apocalyptic tradition presented a model narrative of the
future, which had been standardised by the eighth century and which has
been continuously updated ever since.

The apparent monotony of this chronological sequence of events was
transcended by typological reasoning. Byzantine apocalypses structure

the history of the future along typological patterns that continue the historical interconnections between Old and New Testament characters and events. The typology of the future introduces nonlinear interweavings that turn the incomplex arrow of time into an entangled clothesline that accommodates the intermittent nature of specific providential events. The reader/listener is, thus, transferred from his or her contemporary world of, for instance, the middle Byzantine period back to the antediluvian age, or to the period of the Egyptian captivity, or to the days of the Incarnation, while at the same time picturing the events to come. He/she becomes a spiritual time traveller into the past and back into the future. Consequently, it can be asserted that typological eschatology does not only belong to Biblical Studies (Goppelt 1966) or to the Joachimite apocalypticism of the High Middle Ages (Schmidt-Biggemann 2004: 381–392); it also belongs to Byzantium. Byzantine apocalyptists continuously used and reused typological schemes, which probed and prescribed, much as in dress rehearsals, an orderly transition into a post-apocalyptic world.

Bibliography

Aerts, Willem J. 2010. Hagar in the So-Called Daniel-Diegesis and in Other Byzantine Writings. In *Abraham, the Nations, and the Hagarites: Jewish, Christian, and Islamic Perspectives on Kinship with Abraham*, eds. Martin Goodman, George H. van Kooten, and Jacques T.A.G.M. van Ruiten, 465–474. Leiden: Brill.

Aerts, Willem J., and George A.A. Kortekaas, eds. 1998. *Die Apokalypse des Pseudo-Methodius: Die ältesten griechischen und lateinischen Übersetzungen*. Corpus Scriptorum Christianorum Orientalium, vol. 569. Leuven: Peeters.

Alexander, Paul J. 1962. The Strength of Empire and Capital as Seen Through Byzantine Eyes. *Speculum* 37 (3): 339–357.

———— ed. and trans. 1967. *Oracle of Baalbek: The Tiburtine Sibyl in Greek Dress*. Dumbarton Oaks Studies, vol. 10. Washington, DC: Dumbarton Oaks Center for Byzantine Studies.

————. 1968. Medieval Apocalypses as Historical Sources. *American Historical Review* 73 (4): 997–1018.

Angelov, Dimiter. 2007. *Imperial Ideology and Political Thought in Byzantium: 1204–1330*. Cambridge: Cambridge University Press.

Baun, Jane. 2007. *Tales from Another Byzantium: Celestial Journey and Local Community in the Medieval Greek Apocrypha*. Cambridge: Cambridge University Press.

Berger, Klaus, ed. and trans. 1976. *Die griechische Daniel-Diegese*. Leiden: Brill.

Böhlig, Gertrud, ed. 1973. *Ioannis Caminiatae de expugnatione Thessalonicae*. Corpus fontium historiae Byzantinae, vol. 4. Berlin: De Gruyter.

Brandes, Wolfram. 1997. Anastasios ὁ δίκορος: Endzeiterwartung und Kaiserkritik in Byzanz um 500 n. Chr. *Byzantinische Zeitschrift* 90 (1): 24–63.

———. 2007. Konstantinopels Fall im Jahre 1204 und 'apokalyptische' Prophetien. In *Syriac Polemics. Studies in Honour of Gerrit Jan Reinink*, eds. Wout Jacques van Bekkum, Jan Willem Drijvers, and Alexander C. Klugkist, 239–259. Orientalia Lovaniensia Analecta, vol. 170. Leuven: Peeters.

———. 2011. Endzeiterwartung im Jahre 1009 A.D.? In *Konflikt und Bewältigung: die Zerstörung der Grabeskirche zu Jerusalem im Jahre 1009*, ed. Thomas Pratsch, 301–320. Millennium-Studien zu Kultur und Geschichte des ersten Jahrtausends n. Chr., vol. 32. Berlin: De Gruyter.

Constantinou, Eugenia S., trans. 2011. *Andrew of Caesarea. Commentary on the Apocalypse*. The Fathers of the Church: A New Translation, vol. 123. Washington, DC: Catholic University of American Press.

Davidson, Richard M. 2011. The Eschatological Hermeneutic of Biblical Typology. *TheoRhēma* 6 (2): 5–48.

De Jong, Irene J.F., and René Nünlist, eds. 2007. *Time in Ancient Greek Literature: Studies in Ancient Greek Narrative, Volume Two*. Mnemosyne: Bibliotheca Classica Batava, vol. 291. Leiden: Brill.

DiTommaso, Lorenzo. 2005. *The Book of Daniel and the Apocryphal Daniel Literature*. Studia in Veteris Testamenti Pseudepigrapha, vol. 20. Leiden: Brill.

Florovsky, Georges. 1956. Eschatology in the Patristic Age. *The Greek Orthodox Theological Review* 2: 27–40.

Frye, Northrop. 1982. *The Great Code: The Bible and Literature*. New York: Harcourt Brace Jovanovich.

Genette, Gérard. 1980. *Narrative Discourse: An Essay on Method*, trans. Jane E. Lewin. Ithaca, NY: Cornell University Press.

Goppelt, Leonhard. 1964. Apokalyptik und Typologie bei Paulus. *Theologische Literaturzeitung* 89: 321–344.

———. 1966. *Typos: Die typologische Deutung des Alten Testaments im Neuen*. Beiträge zur Förderung christlicher Theologie, ser. 2, vol. 43. Darmstadt: Wissenschaftliche Buchgesellschaft.

Grumel, Venance. 1958. *La Chronologie*. Traité d'Etudes Byzantines, vol. 1. Paris: Presses Universitaires de France.

Halliday, William R. 1913. *Greek Divination: A Study of its Methods and Principles*. London: Macmillan.

Kraft, András. forthcoming. Miracles and Pseudo-Miracles in Byzantine Apocalypses. In *Recognizing Miracles in Antiquity and Beyond*, ed. Maria Gerolemou. Berlin: De Gruyter.

Leone, Pietro A.M., ed. 1968. *Ioannis Tzetzae Historiae*. Pubblicazioni dell'Istituto di filologia classica, vol. 1. Naples: Libreria Scientifica Editrice.

——— ed. 1972. *Ioannis Tzetzae Epistulae*. Leipzig: Teubner.

Magdalino, Paul. 2002. Une prophétie inédite des environs de l'an 965 attribuée à Léon le Philosophe (MS Karakallou 14, f.253r-254r.). *Travaux et Mémoires* (= *Mélanges Gilbert Dagron*) 14: 391–402.

———. 2003. The year 1000 in Byzantium. In *Byzantium in the Year 1000*, ed. Paul Magdalino, 233–270. Brill: Leiden.

———. 2007. Isaac II, Saladin and Venice. In *The Expansion of Orthodox Europe: Byzantium, the Balkans and Russia*, ed. Jonathan Shepard, 93–106. Aldershot: Ashgate Variorum.

Maisano, Riccardo. 1975. *L'Apocalisse apocrifa di Leone di Constantinopoli*. Naples: Morano.

Mango, Cyril. 1980. *Byzantium: The Empire of New Rome*. New York: Charles Scribner's Sons.

———. 1984. Le temps dans les commentaries byzantins de l'Apocalypse. In *Le temps chrétien de la fin de l'Antiquité au Moyen Âge: IIIe-XIIIe siècles, Paris 9–12 mars 1981*, ed. Jean-Marie Leroux, 431–438. Colloques internationaux du CNRS, vol. 604. Paris: Éditions du CNRS.

McGinn, Bernard. 1978. Angel Pope and Papal Antichrist. *Church History* 47 (2): 155–173.

Petit, Louis, Martin Jugie, and Xenophon A. Sidéridès, eds. 1935. *Œuvres complètes de Gennade Scholarios*, Vol. 4. Paris: Maison de la Bonne Press.

Podskalsky, Gerhard. 1974. Marginalien zur Byzantinischen Reichseschatologie. *Byzantinische Zeitschrift* 67 (2): 351–358.

Pratsch, Thomas. 2005. *Der hagiographische Topos. Griechische Heiligenviten in mittelbyzantinischer Zeit*. Millennium Studies, vol. 6. Berlin: De Gruyter.

Prince, Gerald. 1982. *Narratology: The Form and Functioning of Narrative*. Berlin: Mouton.

Reinink, Gerrit J., ed. 1993. *Die Syrische Apokalypse des Pseudo-Methodius. Corpus Scriptorum Christianorum Orientalium*, vol. 540. Leuven: Peeters.

Rydén, Lennart, ed. and trans. 1974. The Andreas Salos Apocalypse: Greek Text, Translation, and Commentary. *Dumbarton Oaks Papers* 28: 197–261.

———, ed. and trans. 1995. *The Life of St. Andrew the Fool, II: Text, Translation and Notes, Appendices*. Studia Byzantina Upsaliensia, vol. 4/2. Uppsala: Almqvist & Wiksell International.

Ševčenko, Ihor. 2002. Unpublished Byzantine texts on the end of the world about the year 1000 AD. *Travaux et mémoires* (= *Mélanges Gilbert Dagron*) 14: 561–578.

Schmid, Josef, ed. 1955. *Der Apokalypse-Kommentar des Andreas von Kaisareia*, Vol. 1. Munich: Karl Zink Verlag.

Schmidt-Biggemann, Wilhelm. 2004. *Philosophia Perennis: Historical Outlines of Western Spirituality in Ancient, Medieval and Early Modern Thought.* International Archives of the History of Ideas, vol. 189. Dordrecht: Springer.

Schmoldt, Hans. 1972. *Die Schrift "Vom jungen Daniel" und "Daniels letzte Vision". Herausgabe und Interpretation zweier apokalyptischer Texte. PhD dissertation*, University of Hamburg.

Soustal, Peter. 1991. *Tabula Imperii Byzantini. Band 6: Thrakien (Thrakē, Rodopē und Haimimontos).* Philosophisch-historische Klasse, Denkschriften, vol. 221. Wien: Verlag der Österreichischen Akademie der Wissenschaften.

Suggit, John N., trans. 2006. *Oecumenius. Commentary on the Apocalypse.* The Fathers of the Church: A New Translation, vol. 112. Washington, DC: Catholic University of American Press.

Treadgold, Warren. 2004. The Prophecies of the Patriarch Methodius. *Revue des Études Byzantines* 62: 229–237.

Turner, Christopher J.G. 1964. Pages from the Late Byzantine Philosophy of History. *Byzantinische Zeitschrift* 57 (2): 346–373.

Van Dieten, Jan-Louis, ed. 1972. *Nicetae Choniatae Orationes et Epistulae.* Corpus Fontium Historiae Byzantinae, Series Berolinensis, vol. 3. Berlin: De Gruyter.

Van Donzel, Emeri, and Andrea Schmidt, eds. 2009. *Gog and Magog in Early Eastern Christian and Islamic Sources: Sallam's Quest for Alexander's Wall.* Brill's Inner Asian Library, vol. 22. Leiden: Brill.

Vasiliev, Alexander A., ed. 1893. *Anecdota Graeca-Byzantina.* Moscow: Imperial University Press.

Westerink, Leendert G. 1975. Nicetas the Paphlagonian on the End of the World. In *Essays in Memory of Basil Laourdas*, ed. Leendert G. Westerink, 177–195. Thessaloniki: E. Sfakianakis & Sons.

Zervos, George T. 1983. The Apocalypse of Daniel (Ninth Century A.D.): A New Translation an Interpretation. In *The Old Testament Pseudepigrapha I: Apocalyptic Literature and Testaments*, ed. James H. Charlesworth, 755–770. Garden City, NY: Doubleday & Company.

CHAPTER 5

Unknown or Uncertain? Astrologers, the Church, and the Future in the Late Middle Ages

Klaus Oschema

Astrology enjoys an ambivalent status as a means to gather information about the future—and this does not only refer to the sceptical, if not overtly inimical, position held by the majority in contemporary Western societies.[1] In fact, controversial debates about and ambiguous attitudes towards astrology do have quite a long history. In medieval Europe, for example, many individuals either had a keen interest in astrology or they, at least frequently, sought advice regarding future events, which they hoped could be furnished, amongst others, by astrologers. At the same time, a highly critical discourse existed, the protagonists of which

---

[1] For some indicative numbers concerning the lasting popularity of horoscopes and other astrological prognostications, cf. Minois (1996: 565–566). Even though astrology continues to thrive, see Esquerre (2013), only a minority (albeit a significant one) accepts it as a 'science'; see Allum (2011).

K. Oschema (✉)
Department of History, Ruhr-Universität Bochum,
Bochum, Germany

© The Author(s) 2017
S. Baumbach et al. (eds.), *The Fascination with Unknown Time*,
DOI 10.1007/978-3-319-66438-5_5

were mainly clerics.[2] In spite of the resulting vivid arguments, and even though many people in pre-modern Europe considered astrology to be of great importance,[3] the very object of astrology and its role in society has long been neglected by modern historians of the European Middle Ages.[4]

This reluctance of modern historians to engage with this subject can at least partly be explained by the deliberately 'enlightened' approaches that characterised Historical Studies and Medieval History in the period of their establishment as scientific disciplines in their own right.[5] In comparison with their predecessors, the academic historians of the nineteenth and early twentieth centuries described past events less to illustrate the greatness of God's creation and His influence in this world: following the ideal of a strictly scientific approach, many of them merely wanted, as Ranke's famous dictum says, to "show what had happened" (Moeglin 2015: 7 [my translation]). In spite of this claim to a newly established objectivity, the very same historians could of course not free themselves from influences that one might call ideological: although the new 'scientific' history was no longer written to prove God's greatness, much of it now served the interests of the new ideal of the nation state.[6]

This aspect cannot be discussed in more detail here because it would lead to an entirely different subject. It has to be noted, however, because it explains the reluctance of modern historians to tackle matters concerning astrology: in the wake of the same enlightenment thinkers whose work determined the new character of history itself, the practice of astrology was rapidly dismissed as superstitious and futile (but now on the basis of secular, scientific thought). If history was to serve the interests of the nation state, it had to consider hard facts and identifiable events, and not analyse dubious belief systems that were of no value anyway. In a quite paradoxical manner, one might argue, this attitude led to a strange coalition of enlightened historians and the Church, because the

---

[2] See, for example, Boudet (2006), Wedel (1920).

[3] Cf. Garin (1983).

[4] A noticeable exception is furnished by the monumental work of Thorndike (1923–1958). For a brief overview on the research tradition, see Mentgen (2005: 1–9).

[5] For history's new role from the early nineteenth century onwards, see Moeglin (2015: 13–14).

[6] See, for example, Berger and Lorenz (2010).

latter had always kept its distance from astrology, if it was not outright inimical.

Although the study of astrology and its role in pre-modern Europe and other regions has intensified considerably over the past decades,[7] the fundamental attitude and beliefs that lay behind this strange coalition can provide a useful point of departure. In this chapter I want to demonstrate that the concept of 'fascination' as outlined in the introduction to this volume indeed constitutes a useful tool that can help us better understand the attitudes of medieval individuals towards time.[8] To do so, I focus on practices that helped to gather knowledge about the future (in the eyes of the practitioners and their clients). This particular focus on the future is based on the idea that the ambivalent phenomenon of fascination is somehow connected with the attractive force of the 'unknown' and that consequently there seems to be a fundamental difference between the past and the future: both dimensions were—and are, of course—inaccessible to human beings, who inevitably live and act in their own present.[9] But although the past certainly interested medieval thinkers, as is attested by the importance of historiographical writing, a lack of knowledge merely meant that no texts transmitted specific information on events that could theoretically be known without further difficulty. The future, however, poses an entirely different problem, namely an ontological one: as it has not yet taken place, different events can develop, different futures can become reality.[10]

To demonstrate the ensuing effects, I focus on phenomena connected with the practice of astrology and show how the sheer existence of these practices suggests the fascination of an astrologer's client with time, or, to be more exact, with the future and the things to come, in a way that

---

[7] A critical survey on recent publications is a desideratum. Apart from Boudet (2006), recent contributions include Azzolini (2013), and several collective volumes in the series Micrologus' Library as well as from the International Consortium for Research in the Humanities "Fate, Freedom and Prognostication" at the University of Erlangen-Nuremberg. See most recently Deimann and Juste (2015).

[8] One has to keep in mind, however, that the analytical notion of 'fascination' must not be confused with the medieval use of Latin *fascinatio*: When, for example, Roger Bacon uses the word in the thirteenth century, it clearly designates the negative phenomenon of "bewitching," cf. Roger Bacon, *Opus majus*, vol. 1., ed. Bridges, 1897: 143, 398.

[9] On Niklas Luhmann's concept of simultaneity, cf. Landwehr (2012: 29–30).

[10] See Zimmerli (1998: 221–248); cf. below, n. 22.

largely exceeds purely pragmatic approaches. To avoid over-generali-sation and ahistoricity, I highlight the particular difficulties with which medieval astrologers have been confronted and then discuss examples that demonstrate the strategies that late medieval astrologers developed to satisfy both the exigencies of the church and their own need to adver-tise their art.

MEDIEVAL THINKING AND THE FUTURE: A VERY SHORT INTRODUCTION

In contrast to modern societies with their secularised concepts of time and history,[11] the Latin Christian inhabitants of medieval Europe had a quite precise idea of how things would end[12]: The Bible furnished help-ful, reliable, and authoritative information on the end of the secular world and on the ensuing eternal rule of God.[13] It also gave the indi-vidual an idea of what to prepare for, including prescriptions and rela-tively concrete advice.[14] After all, the material world only constituted a transitory phase—every Christian's true home being God's kingdom, an individual's existence on this earth only represented a kind of pilgrimage in an exile.

Based on these characteristics, many scholars of medieval Europe would argue that medieval thinkers did not develop an autonomous idea of a secular future that would have been worth deeper reflection. This idea harmonises with an observation by František Graus, who showed that medieval 'utopian' texts did not place their imagined ide-als in a different time but rather in distant places (although the idea of a lost 'Golden Age' existed) (1967: 8–9). For medieval thinkers, time only spans the distance between the extreme poles of God's creation and His eternal kingdom after the events of the Apocalypse. Faced with the fundamental importance of these extremities, reflections on time on a

---

[11] Cf. especially the work of Koselleck (2004).

[12] See, amongst others, Minois (1996) and Schmieder (2015).

[13] This does not only concern the events of the Apocalypse but also preceding devel-opments, as is attested, for example, by the works of Joachim of Fiore: His prophetical interpretations were based on interpretation on the parallelism of events and figures in the Old and New Testament with the time after Christ. See, for example, Reeves (1999), Patschovsky (2003).

[14] Cf. Schmieder (2015).

smaller scale and on the events that it would bring in one's own lifetime could not really matter—or could they?

Of course, things are never quite as clear as they seem to be at first glance, and recent research does in fact challenge this well-established picture. Even some canonical texts can demonstrate that time was clearly perceived to be ambivalent—and thus fascinating?—by late antique and medieval Christian thinkers. One of the most famous examples is furnished by St. Augustine, who formulated what was to become a topos in reflections about time: "What then is time? If no one asks me, I know what it is. If I wish to explain it to him who asks, I do not know." (Saint Augustine, *Confessions*, trans. Watts, 1912: 238–239). Concerning the future, the church father explicitly continued that "if there was nothing to come, there would be no time to come" (*non esset futurum tempus*), only to conclude that the future remained elusive in the sense that it "is not yet" (*iam non est*) (ibid.).[15] Another reflection of the difficulty in coming to terms with the nature of time can be found in Honorius Augustodunensis's twelfth-century *Imago mundi*, a brief encyclopaedic text, that gives the somewhat obscure and metaphorical definition: "[Secular] Time is the shadow of eternity" (Honorius Augustodunensis, *Imago mundi*, ed. Flint: 92 (II.3) [my translation]).[16]

Dark and figurative notions such as these literally invited symbolic approaches and prophetic ideas. As is shown by Anke Holdenried, András Kraft, and Christian Hoffarth in the present volume, one means to gather information on the future was furnished by prophecy and the interpretation of biblical writings. In addition to the picture gleaned from the divinely inspired practice of prophecy and the religious reasoning of biblical exegesis, a closer look at the practice of astrologers can furnish another approach inviting us to modify well-established paradigms.

In the early fifteenth century the Dominican Laurens Pignon, who was later to become the confessor of the Burgundian duke Philip the Good, wrote a treatise in which he fervently argued against the superstitious practice of consulting *devineurs* and which he addressed to John the Fearless, Philip's father and predecessor. Pignon applied this

---

[15] Cf. Schmitt (2000: 5).

[16] "Tempus autem mundi est umbra ęvi."

name (i.e. *devineurs*) to individuals who claimed to be able to foretell
the future by a variety of means, including astrology (Laurens Pignon,
Contre les devineurs, ed. Veenstra 1998: 230).[17]

If we are to believe Pignon, he wrote his treatise because he had seen
several "famous and powerful lords" consult and believe this kind of
predictors (much to his frustration, one is tempted to add) (ibid. 225
(Prol.) [my translation]). Although it is difficult to judge the real importance of this phenomenon in quantitative terms,[18] two arguments under-
line its relevance in Pignon's time: First, chronicles as well as accounting
attest to the presence of astrologers and other kinds of 'experts' of the
future at late medieval courts (albeit with chronological and regional variations).[19] Second, the sheer fact that Pignon took the trouble to compile a text that covers more than 100 pages in its modern edition is quite
telling.

Being a Dominican, Pignon worried about the theological dangers
that are connected with any attempt to predict the future, and forecasts by astrological means occupied a special place in his argument
(I will come back to the reasons). What seems particularly important for
the question of the fascination of astrologers' clients with time, however,
are the reasons he gives to explain the alleged success of the *devineurs*:
according to Pignon, they pretended to be able to predict "all the good
and bad fortunes" that their clients' future held for them (ibid. 223
(Prol.) [my translation]).[20]

Although the *devineurs'* activity concerned not only information about future events but also about distant and hidden objects,[21]
the author's main concern clearly is with attempts to gather insight
into the future. This focus is demonstrated not only by the bulk of the

---

[17]On John the Fearless and his court, see Schnerb (2005); for the context of Pignon's
activities see also Véronèse (2001), and ibid. (2008: 413–415 and 422–426).

[18]Veenstra gives a list of (real and probably also fictitious) astrologers who can be found
in the context of the ducal court of Burgundy (1998: 128–134).

[19]See, for example, Veenstra (1998).

[20]"toutes ses fortunes et infortunes qui li sont a avenir".

[21]"Disent appres et donnent a entendre par leur presumptueuse assercion, que il puelent
savoir par la science d'astronomie les traitiés et coses secretes qui se font en divers lieus et
lointain païs" (Laurens Pignon, *Contre les devineurs*, ed. Veenstra, 1998: 223 (Prol.)). For
late medieval and early modern obsessions with hidden treasures and their revelation in
dreams, cf. Poley (2015: 93–104) and Dillinger (2015: 105–125).

arguments he develops—which, by the way, rely heavily on the pertinent reasoning Thomas Aquinas already presented in his *Summa Theologica* (Veenstra 1998: 214–222). Pignon actually makes this point quite explicit, when he explains that "every divination primarily seeks knowledge about the things to come" (Laurens Pignon, *Contre les devineurs*, ed. Veenstra 1998: 236 (I.3) [my translation]).[22] Now, as some of these events to come—and theologically the most important—have actually been revealed to the community of believers in the book of Revelation, these are clearly not the phenomena in which astrologers' clients seem to have been interested.[23] Although Pignon's text only provides indirect evidence, it implies that individuals who consulted *devineurs* were mainly concerned with very secular questions (which he mentions explicitly), including their own health as well as the choice of advantageous moments for specific actions such as beginning a pilgrimage. As far as medical decisions are concerned, Pignon actually approves the consultation of astrologers (ibid., 233 (I.2)).[24]

Beyond this circumscribed context, however, Pignon positioned himself as a determined adversary of astrological practices who denounced the all-too-apparent attraction of knowledge about the future. Of course, he was not the only author who showed this attitude. To cite only one of the most famous near-contemporary forerunners, in the 1370s the French cleric and scholar Nicole Oresme wrote about "this sickness, this miserable love and disastrous desire to know future fates" (Nicole Oresme, *Livre*, ed. Coopland 1952: 78 (c.9) [my translation]).[25]

---

[22] "Toute divinacion tent principalment a avoir congnoisance des coses a advenir." The French noun *avenir* appears in the Late Middle Ages; Latin *futurum* had mostly been used in the plural (*futura*), possibly implying the idea of a potential openness, see Schmitt (2000: 5).

[23] Nicole Oresme explicitly declared in his *Livre de divinacions* that these were the things actually worth knowing—and that the book of Revelation furnished a legitimate means to acquire this knowledge, cf. Nicole Oresme, *Livre de divinacions*, ed. Coopland, 1952: 114 (c. 16). On Oresme see most recently (with further bibliographical references) Nothaft (2016: 281–283).

[24] "Et est bien cose convenable et honneste que les segneurs ayent en leur service telz astronomiens ayans collacion avec leur phis<i>ciens."

[25] "ceste maladie, ceste miserable amour, et maleureux desir de savoir les fortunes avenir." I slightly adjusted Coopland's translation; the cited phrase is a quote from Statius's *Thebaid* (III 550–565). Oresme's polemic has to be interpreted in the context of the dynamics between factions at the court of Charles V, see Grant (1997).

Similar to Pignon, Oresme positioned himself as a fervent adversary of 'judicial' astrology, that is, those astrological practices which tried to predict in detail future events that depended on human actions and decisions—or what he calls "effects of fortune" (*effets de fortune*) (ibid., 56 (c.2); trans. ibid. 57).

To do justice to late medieval attitudes towards astrology, however, one has to keep in mind that even polemical authors usually distinguished quite clearly between several branches of the science, which they called quite indiscriminately either 'astrology' or 'astronomy' (Vanden Broecke 2003: 7–27). In their classification, observations of the stars were not only licit but even laudable, because they inspired a sense of the greatness of God's creation. In addition, even some practices that modern authors would categorise as 'astrological' were allowed, namely, a broad range of approaches that relied on phenomena which were interpreted as natural effects of astral influence. The latter included mainly the world of the material such as the weather or the human body. In this sense, astrology could legitimately be used to predict events such as rain, storms, or temperature changes on the one hand and to choose adequate medical treatment on the other hand. As the microcosm of the human body was thought to be related to the macrocosm, astral constellations were believed to have an effect on dietary advice and also on the choice of such medical procedures as the widespread practice of bloodletting (Akasoy et al. 2008).

All these fields directly concerned the future on a scale that was far from the eschatological—and astrology was recognised as a legitimate or even necessary auxiliary science: a physician was supposed to be well versed in astrology. In fact, the necessity of pertinent knowledge even led to the inclusion of lectures on astrology in the curriculum of medical faculties in late medieval universities (Park 1985: 59; Grant 1996: 137). In the perspective of a thinker like Oresme, this branch of astrological practice could be resumed under the heading of "effects of nature" (*effets de nature*) (Nicole Oresme, *Livre*, 1952: 56 (c.2)).

All became more difficult, however, when human decisions and actions came into play: although stellar effects on material objects could easily be imagined and explained, the idea of astral influence on the more noble matter of the human soul or reason would have seemed at once unthinkable and inappropriate. What is more, the idea that human decision making and actions could actually be foretold by looking at the stars (which might function as signs or causal effects) implied that the individual had no actual say in the matter; either a person's actions would be predetermined or the attempt to predict them was valueless. But the idea of predetermined actions implied serious ethical and theological consequences, because how could any individual be held responsible for his or her deeds if these were not decided upon by free will? This argument was purely based on logical reasoning; it had already been made by St. Augustine, and medieval theologians did not cease to repeat it.[26]

However, late-medieval authors did not always follow the logical implications of their arguments to the end: according to Pignon, for example, God's relationship to time differed from the human perspective. Men could not see into the future, as far as human actions were concerned, but this was not true for God: He could freely dispose of present and future events, which were available to Him in a kind of "simultaneous" presence (Laurens Pignon, *Contre les devineurs*, 1998: 334 (III.6), 322 (III.1)). Although this might constitute a convincing argument in the sense that it reinforced the image of an omniscient and omnipotent God, Pignon did not discuss any further the question of how this preordained future that existed for God could be harmonised with the open future experienced by human beings.

ASTROLOGERS AND THE FUTURE: CLAIMS AND LIMITATIONS

Much else could, of course, be said about medieval thinking about the future[27] as well as about the arguments theologians brought forth against judicial astrology. Amongst other aspects, it would be interesting to discuss in more detail the role of empirical proof in the polemics of theologians, who were clearly not only interested in theological reasoning but also well versed in the practice and limitations of astrological prognostications.[28]

---

[26]Cf., for example, Laurens Pignon, *Contre les devineurs*, 1998: 311 (II.5); Nicole Oresme, *Livre*, 1952: 90 gives a weaker version of the argument.

[27]Cf. Minois (1996), Burrow and Wei (2000).

[28]Apart from Nicole Oresme, *Livre de divinacions*, 1952 and Nicole Oresme, *Quaestio*, ed. Caroti, 1976, one might refer to his contemporary Henry of Langenstein, see Pruckner (1933) and Nothaft (2016: 283–284). For the interesting case of Pierre d'Ailly, who changed his attitude towards astrology profoundly, see Smoller (1994); cf. Millet and Maillard-Luypaert (2015: 249–260).

In the present context, however, I want to concentrate on astrologers' reactions towards the polemics with which they were confronted and on the way that these reactions reflect the fascination with the future. As a considerable number of astrologers enjoyed successful careers in different positions at court, in cities, and at universities, it seems only natural to ask what they had to offer to their clients: what were their claims and what strategies did they develop to avoid the accusation of illicit (or even heretical) practices? The latter aspect especially must have been of increasing importance to them, because many late medieval adversaries of astrology did not hesitate to stigmatise any method and claim of foretelling the future as an illegitimate advance into a domain that was solely reserved to God. If someone claimed to have insights into this domain, he or she was thus either a fraud or had to have had help from demonic beings.[29]

From an analytical perspective, one should first note that astrological prognostications did not necessarily have to give adequate information about the future to be efficient. As has been argued convincingly by Georges Minois and Monica Azzolini, many texts that purported to talk about the future actually first and foremost addressed their present audience—either to become self-fulfilling or self-destroying prophecies (Minois 1996: 12–14, 577) or to function as arguments in political quarrels and polemics (Azzolini 2010: 135–145; Azzolini 2013). This instrumentalisation could be performed in several ways, as was noted by Oresme in the late fourteenth century: He distinguished a variety of pseudo-prognostications, including the practice to artificially arrange for the predicted events to come true by means of fraud or magic (Nicole Oresme, *Livre*, 1952: 92–102 (cap. 12)).

Contemporary authors thus reflected upon fraudulent techniques, but they were less occupied with the social and political uses of false prognostications. Accordingly, astrologers who tried to justify their practice and to explain its value—which they did quite frequently in short

---

[29] Boudet speaks of the "uncertain frontier between astrology and magic" (2006: 511 [my translation]). In fact, some authors declared any attempt to gain insight into a field that was reserved to God illicit, cf. Laurens Pignon, *Contre les devineurs*, ed. Veenstra, 1998: 274 (II.2.1): "Mais voloir savoir les cozes secretes et <a> avenir par industrie et engin humain, c'est voloir monter plus haut que creature ne doit et userper le droit de Dieu auquel seulement telle coze apartient, non a autre, se Dieu par grace ne li veult reveler et magnifester".

treatises with annual predictions, the so-called *Judicia anni*[30]—rather underlined the advantages of knowing about the future. This strategy also had a longer tradition: in the mid-thirteenth century Roger Bacon declared that astrologers, although they could not immediately predict individual human action, were able to identify motives and inclinations of humans that were influenced by the stars (Roger Bacon, *Opus Majus*, vol. 1, ed. Bridges 1897: 249).[31] According to his argument, this kind of knowledge could then be used in favour of the Christian religion, as it allowed making people, cities, and regions better—in short, it could help to ameliorate the future (ibid. 402).[32]

What Bacon had in mind here largely consisted in preparedness and appropriate reactions to environmental influence. Although he certainly was an extraordinary figure, it would be erroneous to simply dismiss his ideas as marginal. Quite on the contrary, the concept of preparedness for the things to come, to handle them better, frequently appears in later justifications of astrology. When the physician and astrologer Conrad Heingarter wrote a prognostication for his prince John, duke of Bourbon, in 1477, he echoed Ptolemy's "Tetrabiblos" in underlining that "future evil, if it is known, can be avoided or at least reduced, especially by good advice" (Conrad Heingarter, "Prognostication" [1477]: fol. 1r [my translation])[33]; unfortunately, he did not specify whether he referred to purely natural phenomena (such as meteorological influence) or if he meant to include human behaviour.

---

[30] On the *Judicia anni*, see below; they often contained a brief justification of the astrologers' art; cf. Bauer (1994: 167–168) on prognostications in the German vernacular. From ca. 1490 onwards, the production of vernacular texts, most frequently called *Practica*, superseded the number of Latin treatises (which equally attests to the existence of lay market), see Green (2012: 112, 123) (fig.).

[31] On Bacon and astrology, cf. Power (2011: 63–78), Hackett (1997: 175–198).

[32] "Si igitur Christiani scirent haec opera auctoritate papali facienda ad impedienda mala Christianorum, satis esset laudabile, et non solum propter mala repellenda, sed ad promotionem quorumcunque utilium. Et quia personae, et civitates, et regiones secundum praedicta possunt alterari in melius, et ut vita quantum sufficit prolongetur, et omnes res utiliter procurari, atque multo majora fieri, quam praesenti scripturae debeant commendari, non solum in naturalibus, sed in moralibus scientiis et artibus, sicut patuit per Moysen et Aristotelem".

[33] "... et si futurum malum presciatur notabiliter per optimum concilium tolli aut saltem mitigari potest." A detailed biography of Heingarter is a desideratum; see Préaud (1984: 71–94, 141–155).

Heingarter's argument convincingly justified astrological advice on the future; however, it relied on the condition that the astrologer's predictions were trustworthy. Trivial as this might seem, what has been said about the late medieval attitude towards astrology in the first part of this chapter makes clear that such a claim was prone to provoke conflict. Astrologers were in a delicate position: on the one hand, they basically had to assert that their art furnished reliable predictions on future events and developments—what would otherwise be the point of consulting them? On the other hand, this very claim potentially brought them into conflict with church doctrine. Although we know of only a few astrologers who had effectively been accused of being heretics, tried, and executed—and their astrological practices had not been at the heart of the accusations but rather the performance of magical practices that implied contact with demons or the devil himself[34]—the danger was real, especially when astrologers were not willing to dispense with the advantages of effective promotion of their business.

These effects can be seen in the *Judicia anni* mentioned earlier: written and published in increasing numbers from the fourteenth century onwards (and well into the early modern period),[35] these texts provided their readers with an overview of important events and developments of the year to come. In spite of their broad dissemination, the *Judicia* have mostly been neglected in modern research, even by historians of astrology[36]: In contrast to individual nativities ('birth horoscopes'),[37] they seem less promising for detailed historical contextualisation; at the same time, their level of scientific sophistication cannot compete with elaborate treatises on the theory and practice of astrology. *Judicia anni* thus were something like the 'pulp literature' of late medieval astrological expertise—interesting for a limited period of time and written in a somewhat standardised manner, they were perhaps even an annoying task for some of their authors. At Bologna, the annual redaction and publication

---

[34] In 1327 Cecco d'Ascoli was burned at Florence but mainly for having invoked demons, see Boudet (2006: 401); cf. Rigon (2007). On Jean de Bar, who was burned in Paris in 1398, see Veenstra (1998: 67–69).

[35] For the years 1464–1500, see Thorndike (1923–1958, vol. IV: 438–484).

[36] An exception is the (unpublished) M.A. thesis by Tur (2014). Tur currently prepares a Ph.D. thesis on this subject at the Université d'Orléans. See also Contamine (1985); for the sixteenth century, see Barnes (2016).

[37] For a brief introduction, see Smoller (1994: 17).

of a *Judicium* was part of the duties of the university's professors of astrology (Azzolini 2010: 140–141).[38]

On the other hand, these short treatises apparently had considerable success, as is attested by the impressive number of manuscript *Judicia* that have been conserved but also by the fact that they appeared in print from the 1470s onwards. Especially the latter fact indicates that their authors—and their printers—hoped for an audience that was willing to buy them. Although more detailed research is still necessary,[39] we can thus assume that there existed a veritable market for them.

To compete successfully in this developing sector, authors had to satisfy a series of conditions, which can mostly be subsumed under the headings of 'relevance' and 'entertainment': although the latter aspect certainly had an impact on the choice of details and stylistic traits, the former immediately concerned the question of justification. To appear relevant, the text had to discuss subjects that interested the audience: this concerned, amongst other things, the regional scope of the predictions but also the social groups that were mentioned. Although *Judicia anni* soon became relatively standardised and regularly included prognostications on the popes, the Church, and the clergy, or on the Emperor, individual kings, and their realms, other elements show considerable flexibility: Hence authors writing in Italy gave predictions for a series of individual cities (or rather city-states), while these did not matter in *Judicia* from England. Similar effects might have determined the choice of social groups: we may thus hypothesise that Marcus Scribanarius included a section on the "scholars of all faculties" in his "Judicium" on the year 1480 not so much because he considered this group eminently important but because he wanted to address them as his potential readership (Marcus Scribanarius, "Judicium anni 1480": fol. 107r [my translation]).[40]

Although both strategies—regional and social focus—appear to be quite obvious, other aspects of the *Judicia* warrant more detailed analysis. Given the political circumstances, it is hardly surprising that a huge number of texts include predictions on the fate of 'Christianity' and

---

[38] Cf. Juste (2011: 33).

[39] See, however, Green (2012).

[40] On the manuscript—an autograph collection of *Judicia anni* by Hartmann Schedel—see Tur (2015).

the non-Christians.[41] But how to interpret references to the island of Taprobane, probably modern Sri Lanka (Paul of Middelburg, "Judicium anni 1480": fol. 37v)? Even in the fifteenth century, when Europeans already had a more detailed idea of the Asian world, this place seems to be too distant to be of any concrete interest. Its presence could, however, have served to give the text some 'exotic touch' of the kind that attracted late medieval readers, as is witnessed by the overwhelming success of the so-called 'Travels' allegedly written by John of Mandeville.[42] What might be equally important, I believe, is the aura of scholarliness that the author created, and which added to his credibility.

This point leads back to the important aspect of justification: quite frequently *Judicia anni* start out with a (sometimes very) brief introduction into astrological reasoning, explaining that and why their art actually worked. In these crucial passages, which served as a kind of advertisement and justification, astrologers had to circumscribe the scope of their claims. Although these brief sections are far less elaborate than scientific treatises, they can help us understand for which kind of information and which degree of certitude the astrologers' clients might have been hoping. In addition, it is in these passages that the authors outlined the degree of certainty they claimed for their predictions.

As part of their strategy, several authors anticipated potential criticism by insisting on astrology's character as a "divine science" (*divina scientia*) (Marcus Scribanarius, "Super dispositionem anni 1494": fol. 1r). This argument served to prevent negative reactions by critics, all the while enabling the authors to claim a distinguished status: astrology was in second place only after theology. More systematic justifications sometimes chose a more offensive strategy: in his commentary to the *De Sphaira* of Johannes de Sacrobosco, Cecco d'Ascoli explained that several *artes* could provide a certain insight into future events, amongst them pyromancy, hydromancy, necromancy, and geomancy. But these practices merely furnished "some knowledge"; only astrology could provide "true knowledge about the future" (*futurorum cognitio veritatis*) (Cecco d'Ascoli, *Commentary*, ed. Thorndike, 1949: 346

---

[41] For example, an anonymous *Judicium* from Bologna includes a section on "The State of the Arabs, Maurs, and Turks" [my translation], see Anonymous, "Judicium anni 1456," M.S. Cod. 4756, Österreichische Nationalbibliothek: fol. 10r–19v, 13v–14r. On the polemics caused by the Ottoman expansion, cf. most recently Döring (2013).

[42] On this late-medieval 'bestseller,' see, for example, Tzanaki (2003).

[my translation]). This positive appraisal of astrology coincided with John Ashenden's argument in his massive "Summa judicialis de accidentibus mundi" in the mid-fourteenth century and also with the reasoning Matthias of Kemnat developed against prophetical insights into the future in the 1470s. According to Matthias, who vividly decried the lack of mathematical skills at the University at Heidelberg, "the judgments of astrology are more trustworthy and often true" (Matthias of Kemnat, "Epistola astrologica anno 1460": fol. 4v [my translation]).[43]

In addition to this kind of explicit appraisal, the certainty of predictions is even more often implied by simple grammatical form: the narrative passages of most *Judicia* quite plainly talk about things that are 'going to happen,' without any room for ambiguity or reservations. Although this form is undoubtedly adequate as far as astronomical events are concerned, as when Johannes Laet noted for 1479 that there was going to be a lunar opposition on 7 January (Johannes Laet, "Judicium anni 1479": fol. 7r),[44] it seems more far-fetched in other cases. The latter include concrete meteorological predictions—strong winds, clouds, and fog in September—as well as political events, however vague they may be: "This year the King of France or of the house of France will rise strong with abundant force and call to the arms ..." (ibid., 15v [my translation]).[45] As Johannes notes himself at the end of this paragraph, this constitutes a 'particular' prediction (ibid.), which means that it transgresses the limits of the licit 'general' predictions.

In spite of the risks that included not only severe reactions from the Church but also the possibility of being disproved, many authors of *Judicia* followed similar strategies to furnish interesting and potentially relevant advice. At the same time, they employed several techniques to ensure their safety. On the one hand, they skilfully managed to keep most of their predictions on a level that allowed for disavowal or explanation if the predicted events failed to produce themselves.

Even more interesting, on the other hand, is the fact that more than one astrologer counterbalanced the risks with explicit safeguard clauses:

---

[43] "Et sic inde iudicia vere astrologie probabilia sunt et sepe vera." Cf. John Ashenden, "Summa judicialis" [1489]: fol. 1r.

[44] "in januario die vii ... erit oppositio lune."

[45] "Rex Francie seu francigene se hec anno potenter et abundanter eriget et disponet ad arma ...".

At Oxford, one of the most famous astrologers of fourteenth-century England, the aforementioned John Ashenden, quite boastfully reminded his readers in a treatise on the conjunction of Saturn and Mars in the House of Cancer (1357) and a great conjunction of Saturn and Jupiter (1365) of his previous successful predictions.[46] He claimed that he had actually predicted the arrival of a pestilence on the island in a text on another conjunction in 1345. Several years later he came back to this and outright declared that he had proven (*probavi*) this effect (Snedegar 1988, vol. 2: 375),[47] that is, the arrival of the Black Death—a valuable argument for the promotion of his work. At the same time, Ashenden carefully underlined that even his interpretations had limits: the text on the conjunctions of 1357 and 1365 starts out with comments on the signification of the planets involved in the stellar events. Then the author goes on to explain that he wanted to turn to the "particular signification" (*specialius significatio*) of their conjunction—but not without underlining that, in doing so, he had no intention to write anything that "infringed on the Catholic faith" or that "might offend pious ears" (Snedegar 1988, vol. 2: 366 [my translation]). A century later, an unidentified "Petrus Verariensis," who wrote a short treatise on the year 1460, went even a step further when he explained that stellar influence ultimately depended on God: He proposed to analyse and describe "whatever is going to happen and how it is going to happen; but it will not happen where God does not want it to happen" (Petrus Verariensis (?), "Prognostication on 1460": fol. 367r [my translation]).[48]

A better argument could hardly be imagined to navigate between the extreme poles of infringing on God's own prerogative and the desire to know (and write) about the future with the claim to produce certain and valid information. After all, even if the astrologer's prediction proved wrong, he could always argue that God in his wisdom and omnipotence had chosen to let things develop in a way that differed from what

---

[46] The text treated both conjunctions in one work; it was probably written in 1357; cf. Thorndike (1923–1958, vol. III: 338). On Ashenden, see Carey (1992: 73–78, 85–91) and the (unpublished) thesis by Snedegar (1988).

[47] On the validity of Ashenden's claim, cf. Thorndike (1923–1958, vol. III: 326–336) and Carey (1992: 77).

[48] "... quecumque evenient qualitercumque eveniant; non evenient ubi Deus non velit ea evenire." Cf. the brief comment by Thorndike, who erroneously notes that "a year is not given" (1923–1958, vol. IV: 439). See also Juste (2015: 113).

the stars had indicated. Seen from the opposing perspective of Laurens Pignon, however, such a strategy only contributed to the astrologers' fraudulent behaviour: on the one hand, they claimed to provide "certain knowledge" (*cognoissance certaine*) (Laurens Pignon, *Contre les devineurs*, ed. Veenstra 1998: 278 (II.2.3) [my translation]),[49] but on the other hand, they used a broad range of manoeuvres to cover up their mistakes if their predictions failed to come true (ibid., 281 (II.2.3).[50]

CONCLUSION

This short chapter could only hint at some of the problems that surrounded the practice of astrology in the late Middle Ages (as well as in the early modern period) with regard to the general topic of this volume. Focussing on the question of fascination with time, I wanted neither to provide a systematic overview on astrologers' lives and activities nor to produce insights that could contribute to specialised research into late medieval astrology. Instead, I hope to have demonstrated that people in late medieval Europe were actually fascinated with time. Although religious doctrine quite clearly told them that knowledge about the future was the prerogative of God, many people tried to get a glimpse of what the time to come held in store for them. In spite of what is frequently said about the *Weltbild* of the period, this interest was by no means limited to eschatology: the care for one's soul in the face of the Last Judgement certainly had an important role, but people were also interested in knowing about the future on a shorter scale and in a secular sense.

The efforts that went into the practice of astrology provide evidence for this preoccupation. An impressive number of manuscripts and early prints survive and attest to the productivity of the experts of the future, while the accompanying debates and polemics play their part in proving their importance. But what is most impressive, I think, is the considerable effort of astrologers to develop arguments and strategies that allowed them to claim scientific value and certitude for their predictions,[51] all the while trying to avoid open conflict with dogmatic

---

[49] Cf. ibid., 318 (II.6).

[50] Cf. Nicole Oresme, *Livre*, 1952: 92–102 (c. 12).

[51] Medieval astronomy and astrology in fact played a major role in the development of modern science, cf. Grant (1996). On the importance of empirical observations and mathematical skills, see North (2013: 456–484).

positions, which held that no certainty could exist when talking about the future. Although my chapter could only hint at a small choice of writings and arguments, it should have become clear that the involved parties—including a broader audience which read and paid for prognostications—were quite fascinated with time, the future, and its opacity.

Acknowledgments Research for this contribution has profited from my fellowship as "Gerda Henkel Member" at the Institute for Advanced Study (Princeton, NJ) in 2016–2017. I wish to thank the Gerda-Henkel-Foundation and the IAS for their generous support.

BIBLIOGRAPHY

Akasoy, Anna, Charles Burnett, and Ronit Yoeli-Tlalim (eds.). 2008. *Astro-Medicine: Astrology and Medicine, East and West.* Florence: SISMEL.

Allum, Nick. 2011. What Makes Some People Think Astrology Is Scientific? *Science Communication* 33 (3): 341–366.

Anonymous [from Bologna]. n.d. *Judicium anni 1456.* MS Cod. 4756, Fol. 10r–19v. Vienna: Österreichische Nationalbibliothek.

Ashenden, John. 1489. *Summa judicialis de accidentibus mundi.* Venice.

Azzolini, Monica. 2010. The Political Uses of Astrology: Predicting the Illness and Death of Princes, Kings and Popes in the Italian Renaissance. *Studies in History and Philosophy of Biological and Biomedical Sciences* 41: 135–145.

———. 2013. *The Duke and the Stars: Astrology and Politics in Renaissance Milan.* Cambridge: Harvard University Press.

Barnes, Robin B. 2016. *Astrology and Reformation.* New York: Oxford University Press.

Bauer, Barbara. 1994. Sprüche in Prognostiken des 16. Jahrhunderts. In *Kleinstformen der Literatur,* ed. Walter Haug and Burghart Wachinger, 165–204. Tübingen: Niemeyer.

Berger, Stefan, and Chris Lorenz (eds.). 2010. *Nationalizing the Past: Historians as Nation Builders in Modern Europe.* London: Palgrave Macmillan.

Boudet, Jean-Patrice. 2006. *Entre science et nigromance: Astrologie, divination et magie dans l'Occident médiéval, XIIe–XVe siècle.* Paris: Publications de la Sorbonne.

Burrow, John A., and Ian P. Wei (eds.). 2000. *Medieval Futures: Attitudes to the Future in the Middle Ages.* Woodbridge: Boydell & Brewer.

Carey, Hilary M. 1992. *Courting Disaster: Astrology at the English Court and University in the Later Middle Ages.* Basingstoke: Macmillan.

Cecco d'Ascoli. 1949. Commentary. In *The 'Sphere' of Sacrobosco and Its Commentators,* ed. Lynn Thorndike, 343–411. Chicago: The University of Chicago Press.

Conrad Heingarter. 1477. *Prognostication for John of Bourbon.* MS lat. 7447. Paris: Bibliothèque Nationale de France.

Contamine, Philippe. 1985. Les prédictions annuelles astrologiques à la fin du Moyen Âge: genre littéraire et témoin de leur temps. In *Histoire sociale, sensibilités collectives et mentalités. Mélanges Robert Mandrou,* 191–204. Paris: Presses Universitaires de France.

Coopland, George W. 1952. *Nicole Oresme and the Astrologers: A Study of His Livre de Divinacions.* Liverpool: Liverpool University Press.

Deimann, Wiebke, and David Juste (eds.). 2015. *Astrologers and Their Clients in Medieval and Early Modern Europe.* Cologne: Böhlau.

Dillinger, Johannes. 2015. The Good Magicians: Treasure Hunting in Early Modern Germany. In *Everyday Magic in Early Modern Europe,* ed. Kathryn A. Edwards, 105–125. Farnham: Ashgate.

Döring, Karoline. 2013. *Türkenkrieg und Medienwandel im 15. Jahrhundert: Mit einem Katalog der europäischen Türkendrucke bis 1500.* Husum: Matthiesen.

Esquerre, Arnaud. 2013. *Prédire. L'astrologie au XXIᵉ siècle en France.* Paris: Fayard.

Honorius Augustodunensis. 1982. *Imago mundi,* ed. Valerie I.J. Flint. *Archives d'histoire doctrinale et littéraire du Moyen Âge* 57: 7–153.

Garin, Eugenio. 1983. *Astrology in the Renaissance: The Zodiac of Life* [Italian orig. 1976], trans. Carolyn Jackson. London: Routledge & Kegan Paul.

Grant, Edward. 1996. *The Foundations of Modern Science in the Middle Ages: Their Religious, Institutional, and Intellectual Contexts.* Cambridge: Cambridge University Press.

Grant, Edward. 1997. Nicole Oresme, Aristotle's 'On the Heavens', and the Court of Charles V. In *Texts and Contexts in Ancient and Medieval Science. Studies on the Occasion of John E. Murdoch's Seventieth Birthday,* eds. Edith Sylla, and Michael McVaugh, 187–207. Leiden: Brill.

Graus, František. 1967. Social Utopias in the Middle Ages, trans. Bernard Standring. *Past & Present* 38: 3–19.

Green, Jonathan. 2012. *Printing and Prophecy: Prognostication and Media Change, 1450–1550.* Ann Arbor: University of Michigan Press.

Hackett, Jeremiah. 1997. Roger Bacon on Astronomy-Astrology: The Sources of the Scientia Experimentalis. In *Roger Bacon and the Sciences: Commemorative Essays,* ed. Jeremiah Hackett, 175–198. Leiden: Brill.

Johannes Laet. 1479. *Judicium anni 1479.* MS Clm 648, Fol. 1r–18v. Munich: Bayerische Staatsbibliothek.

Juste, David (ed.). 2011. *Les manuscrits astrologiques latins conservés à la Bayerische Staatsbibliothek de Munich.* Paris: CNRS Editions.

——— (ed.). 2015. *Les manuscrits astrologiques latins conservés à la Bibliothèque nationale de France.* Paris: CNRS Editions.

Koselleck, Reinhard. 2004. *Futures Past: On the Semantics of Historical Time*, trans. Keith Tribe. New York: Columbia University Press.

Landwehr, Achim. 2012. Von der 'Gleichzeitigkeit des Ungleichzeitigen'. *Historische Zeitschrift* 295: 1–34.

Laurens Pignon. 1998. Contre les devineurs. In *Magic and Divination at the Courts of Burgundy and France: Text and Context of Laurens Pignon's Contre les Devineurs (1411)*, ed. and trans. Jan R. Veenstra, 205–339. Leiden: Brill.

Marcus Scribanarius. 1494. *Super dispositionem anni 1494*. Rome.

———. 1480. *Judicium anni 1480*. MS Clm 648, Fol. 98r–114r. Munich: Bayerische Staatsbibliothek.

Matthias of Kemnat. 1460. *Epistola astrologica anno 1460*. MS Clm 1817. Munich: Bayerische Staatsbibliothek.

Mentgen, Gerd. 2005. *Astrologie und Öffentlichkeit im Mittelalter*. Stuttgart: Hiersemann.

Millet, Hélène, and Monique Maillard-Luypaert. 2015. *Le Schisme et la pourpre: Le cardinal Pierre d'Ailly, homme de science et de foi*. Paris: Éditions du Cerf.

Minois, Georges. 1996. *Histoire de l'avenir: Des prophètes à la prospective*. Paris: Fayard.

Moeglin, Jean-Marie. 2015. Naissance de la médiévistique? Des antiquaires-érudits aux historiens-professeurs. In *La naissance de la médiévistique: Les historiens et leurs sources en Europe, XIXe–début du XXe siècle*, eds. Isabelle Guyot-Bachy and Jean-Marie Moeglin, 3–31. Geneva: Droz.

Nicole Oresme. 1952. Livre de divinacions. In *Nicole Oresme and the Astrologers: A Study of His Livre de Divinacions*, ed. and trans. George W. Coopland, 50–121. Liverpool: Liverpool University Press.

———. 1976. Quaestio contra divinatores horoscopios, ed. Stefano Caroti. *Archives d'histoire doctrinale et littéraire du Moyen Âge* 43: 201–310.

North, John D. 2013. Astronomy and Astrology. In *The Cambridge History of Science*, vol. 2: *Medieval Science*, ed. David C. Lindberg and Michael H. Shank, 456–484. Cambridge: Cambridge University Press.

Nothaft, C.P.E. 2016. Vanitas Vanitatum et Super Omnia Vanitas: The Astronomer Heinrich Selder, a Newly Discovered Fourteenth-Century Critique of Astrology. *Erudition and the Republic of Letters* 1: 261–304.

Park, Katharine. 1985. *Doctors and Medicine in Early Renaissance Florence*. Princeton: Princeton University Press.

Patschovsky, Alexander (ed.). 2003. *Die Bildwelt der Diagramme Joachims von Fiore*. Ostfildern: Thorbecke.

Paul of Middelburg. 1480. *Judicium anni 1480*. MS Clm 648, Fol. 33r–75r. Munich: Bayerische Staatsbibliothek.

Petrus Verariensis (?). n.d. *Prognostication on 1460*. MS lat. 7336, Fol. 367r–v, 370r–v. Paris: Bibliothèque Nationale de France.

Poley, Jared. 2015. Magic, Dreams, and Money. In *Everyday Magic in Early Modern Europe*, ed. Kathryn A. Edwards, 93–104. Farnham: Ashgate.

Power, Amanda. 2011. The Remedies for Great Danger: Contemporary Appraisals of Roger Bacon's Expertise. In *Knowledge, Discipline and Power: Essays in Honour of David Luscombe*, eds. Joseph Canning, Martial Staub, and E. King, 63–78. Leiden: Brill.

Préaud, Maxime. 1984. *Les astrologues à la fin du Moyen Age*. Paris: J.C. Lattès.

Pruckner, Hubert. 1933. *Studien zu den astrologischen Schriften des Heinrich von Langenstein*. Leipzig: Teubner.

Reeves, Marjorie. 1999. *Joachim of Fiore and the Prophetic Future: A Medieval Study in Historical Thinking*, New and rev. ed. Stroud: Sutton.

Rigon, Antonio (ed.). 2007. *Cecco d'Ascoli: cultura scienza e politica nell'Italia del Trecento*. Rome: Istituto Storico Italiano per il Medio Evo.

Roger Bacon. 1897. *Opus majus*, 3 vols., ed. John H. Bridges. Oxford: Clarendon.

Saint Augustine. 1912. *Confessions*, 2 vols., trans. William Watts. Cambridge: Harvard University Press.

Schmieder, Felicitas (ed.). 2015. *Mittelalterliche Zukunftsgestaltung im Angesicht des Weltendes/Forming the Future Facing the End of the World in the Middle Ages*. Cologne: Böhlau.

Schmitt, Jean-Claude. 2000. Appropriating the Future. In *Medieval Futures: Attitudes to the Future in the Middle Ages*, eds. John A. Burrow and Ian P. Wei, 3–17. Woodbridge: Boydell & Brewer.

Schnerb, Bertrand. 2005. *Jean sans Peur: Le prince meurtrier*. Paris: Payot.

Smoller, Laura A. 1994. *History, Prophecy, and the Stars: The Christian Astrology of Pierre d'Ailly, 1350–1420*. Princeton: Princeton University Press.

Snedegar, Keith V. 1988. John Ashenden and the Scientia Astrorum Mertonensis: With an Edition of Ashenden's 'Pronosticationes', 2 vols. PhD dissertation, University of Oxford, London.

Steven, Vanden Broecke. 2003. *The Limits of Influence: Pico, Louvain, and the Crisis of Renaissance Astrology*. Leiden: Brill.

Thorndike, Lynn. 1923–1958. *A History of Magic and Experimental Science*, 8 vols. New York: Columbia University Press.

——— (ed.). 1949. *The 'Sphere' of Sacrobosco and Its Commentators*. Chicago: The University of Chicago Press.

Tur, Alexandre. 2014. *À l'entrée du Soleil en Bélier: Les prédictions astrologiques annuelles latines dans l'Europe du XVe siècle, 1405–1484*. M.A. thesis, École nationale des chartes. http://theses.enc.sorbonne.fr/2014/tur. Accessed 15 Feb 2016.

———. 2015. Hartmann Schedel, collectionneur et copiste de prédictions astrologiques annuelles. *Bulletin du bibliophile* 278–296.

Tzanaki, Rosemary. 2003. *Mandeville's Medieval Audiences: A Study on the Reception of the Book of Sir John Mandeville, 1371–1550*. Aldershot: Ashgate.

Veenstra, Jan R. 1998. *Magic and Divination at the Courts of Burgundy and France: Text and Context of Laurens Pignon's* Contre les devineurs *(1411)*. Leiden: Brill.

Véronèse, Julien. 2001. Jean sans Peur et la 'fole secte' des devins: enjeux et circonstances de la rédaction du traité Contre les devineurs (1411) de Laurent Pignon. *Médiévales* 40: 113–132.

———. 2008. Contre la divination et la magie à la cour: trois traités adressées à des grands aux XIVe et XVe siècles. *Micrologus* 16 [I saperi nelle corti]: 405–431.

Wedel, Theodore Otto. 1920. *The Mediaeval Attitude Toward Astrology, Particularly in England*. New Haven: Yale University Press.

Zimmerli, Walther Ch. 1998. Zeit als Zukunft. In *Was treibt die Zeit?* vol. 2, *Entwicklung und Herrschaft der Zeit in Wissenschaft, Technik und Religion*, ed. Kurt Weis, 221–248. Munich: dtv.

CHAPTER 6

'From the Unknown to the Known and Backwards:' Representing and Presenting Remote Time in Nineteenth-Century Palaeontology

Marco Tamborini

Fossils are fascinating in and of themselves, as is the history of life that they tell.
(J. John Sepkoski, Jr., What I Did with My Research Career: 144)

Palaeontology is at once a historical and a biological discipline. It analyses and illustrates the biological development of life on Earth over time. Naturalists have always been fascinated with marvellous organisms that were fossilised, for these are manifest remains of a remote and inaccessible past that was presumably dominated by monstrous creatures. From the sixteenth century onwards, naturalists listed and collected these curiosities during their travels and drew catalogues to share them with other savants. In the course of the seventeenth century, several naturalists

M. Tamborini (✉)
Institute of Philosophy, Technische Universität Darmstadt,
Darmstadt, Germany

© The Author(s) 2017
S. Baumbach et al. (eds.), *The Fascination with Unknown Time*,
DOI 10.1007/978-3-319-66438-5_6

started using these catalogues to narrate a natural history that gave access to and control over former time.[1]

Thus, by collecting marvellous fossils, naturalists were always attracted to the structure of a remote and mysterious temporal dimension. This fascination[2] continued throughout the entire eighteenth and nineteenth centuries. For instance, German mining engineer Georg Hartwig summarised the palaeontological fascination with fossils and the unknown in 1863:

> Nature reveals its wonders not only in the unlimited world of stars or in the infinite variety of animals and plants, which adorned and enlivened the surface of earth, but it offers us also a truly amazing wealth of *strangeness and beauty* in the *dark spaces of the subterranean world*. In these spaces, mysterious preservations of fire forces can be found. There, the remains of extinct animals and plants are entombed in successive and consecutive layers. These remains are true medals of creation. *They reveal to geologists' eyes the whole evolution (Entwicklungsgeschichte)* of our planet.[3] (Hartwig, *Unterwelt*, 1863: 3 [my italics])

To Hartwig, strangeness and beauty were merged together in the dark subterranean world, and they revealed an intriguing history to the onlooker. However, as soon as naturalists became aware of the enormous extent of time that made up Earth history, this fascination mutated into a deeper reflection on the epistemic limits of both their research and the data. From this moment onwards, fascination with the past epochs of nature embodied in fossils has both attracted and scared the investigators.

The famous French naturalist Georges-Louis Leclerc, Comte de Buffon (1707–1788), epitomised this antithetic feeling. He was one of the first natural historians to explore the depth of geological time. Buffon was fascinated with the extent of this temporal dimension, which expanded natural history far beyond the capacity of human understanding. At the same time, however, he grasped the destructive role bound to this immense span of time. He did indeed his utmost to investigate the structures of the various epochs of nature to command them:

---

[1] For an in-depth historical analysis of natural history during the sixteenth and seventeenth centuries, see Rudwick (1972, 1985, 2005, 2008).

[2] See introduction for an analysis of this concept.

[3] All translations are mine unless otherwise noted.

As in civil history we consult deeds, seek for coins, or decipher antique inscriptions in order to determine the epochs of human revolutions ... so, in natural history, we must search the archives of the world, recover old monuments from the bowels of the earth, collect their fragmentary remains, and gather into one body of evidence all the signs of physical change which may enable us to look back upon the different ages of nature. (Comte de Buffon, *Histoire naturelle*, 1778: 1)

To create a means that allowed him to better quantify the extent of past time, Buffon developed what has become known as his "steady state theory of earth cooling," according to which "the earth and other planets had all originated simultaneously from a plume of intensely hot material torn from the sun [and] cooled slowly to their present temperatures" (Rudwick 2005: 227). He then experimentally calculated the cooling time of various iron balls from white heat to the stable ambiance and reckoned 74,832 years from the Earth's creation to his days. As a consequence, however, Buffon realised that the amount of time that was literally entombed in the Earth's layers (and its history) threatened to become a sort of dark abyss: "The darkest point [in our theory of the earth] is where we start falling into the abysses of time. There, genius' insight seems not to be enough and, due to the lack of observations, it is powerless to guide us forward" (Comte de Buffon, *Histoire naturelle*, 1778: 44). Thus, Buffon emphasised the ambivalent nature of time in natural history: this vast temporal dimension clearly attracts the scientist, but it also throws him into a dark abyss from which it is quite impossible to escape. Palaeontological time reshapes the human perspective on nature and restricts the human possibility of knowing what happened in the remote past.

Other palaeontologists of the nineteenth century similarly experienced this attraction. British geologist George Julius Poulett Scrope (1797–1876), for instance, emphasised the immense quantity of time that needed to be surveyed to understand Earth's history. The sheer extent of these early periods in Earth's history in turn threatened to narrow "our existence:" we "are in all probability but trifles in the calendar of Nature. It is Geology that, above all other sciences, makes us acquainted with *this important, though humiliating fact*" (Scrope, *Memoir*, 1827: 165 [my italics]). Just like Buffon, Scrope underlined the conflicting feelings caused by reflecting on geological time. Confronted with this vastness, human beings could only be intimidated, even if they were intrigued.

In the course of the nineteenth century, palaeontologists came up with practices that were able to represent and visualise time in a way that conferred a positive value to this dimension in the context of their research. They managed to comprehend unknown time from the known one and the known one from the unknown one. To put it succinctly: "*Method of Geological Investigation* – In whatever way geological history is written, its original investigators have only one method of proceeding– from the known to the unknown–or backwards in the course of time" (Woodward, *Manual*, 1856: 420).

In this chapter I intend to shed light on the nature of the practices that were developed to model and work with palaeontological time. By analysing the epistemic presuppositions that made it possible to proceed from the known to the unknown—and backwards—I furnish insight into central epistemic features of unknown time, thereby providing the basis for a better understanding of the role of fascination and the unknown in historical research. I argue that to understand what the geological past is we must analyse *how* palaeontologists use and work with the records of the past. This examination reveals that to grasp the dark abyss of time palaeontologists substantially modified their visual practices: a shift from palaeontological images as representations of timeframes to palaeontological images as presentations of remote time took place during the nineteenth century.

First, I analyse what can be called the 'destructive' nature of geological time. Geological time destroys its evidence, its traces, and thus its 'deep biological history.' Thus, what happened in the past is underdetermined by the evidence that can potentially be gathered in our present: our knowledge of Earth's deep time and history therefore is always fragmentary (Cleland 2002; Turner 2005).

Second, I investigate how palaeontologists sought to escape from this "dark abyss" that transforms, modifies, and undermines palaeontological knowledge. To not lose their own way in the vastness of time, palaeontologists started trying to understand its immensity by modelling and representing it spatially: this happened through at least two practices that were, *prima facie*, quite different. On the one hand, they identified the fossil record with monuments of the past. This approach enabled them to identify and to visualise specific points in time, which helped infer the immense duration and indetermination of geological time in its entirety. However, this practice could not entirely grasp what happened in the past. Indeed, by equating the fossil record with monuments of the remote past, the palaeontologists could only visualise one individual snapshot at a time of the numerous frames that make up this unknown temporal dimension. On the other hand, by means of tables, charts, and graphs, palaeontologists found another way to proceed from the unknown to the known or backwards. These tools enabled them to visualise and manage unknown time in the form of visual patterns of development. These reconstructed patterns replay 'life's tape,' putting several frames of the past into a unitary sequence.[4] They presented coherent versions of the past.

In my conclusion, I reflect on the epistemic features of 'unknown time' in palaeontology. Palaeontological time can be understood, expressed, and ultimately constructed by means of *spatial presentations* such as tables, graphs, and diagrams or by *pictographic representations* of extinct organisms: both practices share the same constructivist approach, which is also featured in other disciplines such as ethnography.[5] However, visual representations rearrange the characteristics of the fossil record and generate possible data to overcome its fragmentary status. They re-present what already is: they re-present the features of the excavated fossil record. In contrast, presentation is "no longer necessarily focused on copying what already exists—and instead becomes part of a coming-into-existence" (Daston and Galison 2007: 383). Hence, further investigation is required that addresses how the palaeontological representation and presentation of data and their ontology has changed over time.

THE DESTRUCTIVE POWER OF GEOLOGICAL TIME

At the end of the nineteenth century, the prominent American vertebrate palaeontologist William Matthew (1871–1930) underlined the vastness and immensity of palaeontological time by pointing how absolutely tiny man's place in it was: "[Palaeontological] time is measured in geologic epochs and periods, in millions of years instead of centuries. Man, by this measure, is but a creature of yesterday" (Matthew 1915: 9).

---

[4]A thorough analysis of the historical preconditions that made this practice possible can be found in Sepkoski and Tamborini, forthcoming.

[5]For an account of the ways in which ethnography constructs its objects of enquiry to fill, communicate, and display unknown time, see Katja Wehde's chapter in this volume.

This comment was indeed a common trope in the palaeontological and geological literature: since the late eighteenth century, the degraded condition of the human spirit, which emerged from the dark abyss of time, has been made a subject of intense debate (Rossi 1984). This debate culminated with the publication of Charles Darwin's *Origin of Species* (1859) and *The Descent of Man* (1871): Darwin used the immense span of geological time to establish his evolutionary thought (Engels 2008; Kohn 1985; Tamborini 2015b). As a result, the deepness of geological time was seen as a dangerous dimension that could undermine and disintegrate well-rooted ethical and social convictions. Moreover, its immense vastness posed several problems for geological and palaeontological research as well, because it showed the features of the fossil record to be an imperfect and incomplete record of the past, and thus lessened the possibility of establishing palaeontology as a pure biological science.

Intuitively, the fossil record is a material object entombed in rock layers, witnessing the complexity and variety of life forms that appeared on the Earth's surface. Fossils are indeed records of past life. They represent and reflect the features of past epochs. These well-preserved objects that can be admired in natural history museums are not, however, palaeontological data,[6] that is, reliable starting points for biological investigations. Quite the contrary, it is extremely rare to find well-preserved fossils that immediately resemble living organisms and which can be exhibited, used, and taken at face value. In fact, once an organism dies, it is subjected to what have been called taphonomic[7] processes. These processes destroy and change the features of the original organism. For instance, it is extremely rare to find fossils with soft parts preserved and even hard parts, as bones, for instance, are rarely completely preserved. What remains today of many hundreds of dinosaurs in the end amounts to only a few fragmentary bones: palaeontologists have to carefully classify and catalogue all the extracted parts as well as constitute and recreate the missing bones to acquire an adequate picture of these extinct

---

[6] On this definition of data, see Gitelman (2013), Müller-Wille and Charmantier (2012), Massimi (2011), Rheinberger (2011), Cardani and Tamborini (2016), whereas on the notion of data in palaeontology, see Sepkoski and Tamborini, forthcoming, Sepkoski (2013).

[7] First introduced in 1940, taphonomy (from the Greek τάφος [burial] and νόμος (law)) researches the laws behind burial processes (Efremov 1940).

animals. Remote geological time shapes, changes, and destroys the original organism. Thus, the fossil record is always imperfect and incomplete.[8]

Not only are single data, that is, fossils, imperfect and incomplete in palaeontological time, but also time modifies and undermines our knowledge of entire past events. For example, we have little evidence of what wiped out dinosaurs during the Cretaceous–Palaeogene mass extinction (about 65 million years ago) (Keller 2005): in fact, our knowledge of this particular era is particularly fragmentary and incomplete. We have even fewer data about the Permian–Triassic mass extinction (about 252 million years ago), in which nearly 80% of life forms disappeared, or about the so-called Cambrian explosion of life (around 542 million years ago), during which most major animal phyla[9] emerged.

At end of the eighteenth and during the entire nineteenth century, palaeontologists faced these difficulties in investigating ancient time, thus grounding palaeontology as an independent and biological discipline. In a famous passage, the German naturalist Johann Friedrich Blumenbach (1752–1840) wrote, for instance, that the "dead organised bodies are named fossils or petrifacts. They look only more or less like living species" (Blumenbach, *Naturgeschichte*, 1788: 656). The fossil record resembles more or less the extinct organisms and consequently our understanding of the past is incomplete and flawed.

Blumenbach's words set up the classical definition of palaeontological knowledge, as is witnessed by Charles Darwin, who explicitly asserted in the *Origin of Species*:

> For my part ... I look at the natural geological record, as a history of the world imperfectly kept, and written in a changing dialect. ... Of this volume, only here and there a short chapter has been preserved; and of each page, only here and there a few lines. (Darwin, *Origin*, 1964: 310–311)

---

[8]As palaeontologist Michael James Benton nicely put it, "in an ideal world, the best approach to establishing the pattern of the diversification of life would be to collect data from a comprehensive fossil record and to read off the empirical pattern as documented by the fossils. The world, however, is not ideal, and this approach entails many problems of interpretation, not least the quality of the fossil record" (Benton 1997: 490).

[9]In biology a phylum is a principal taxonomic category that ranks above class and below kingdom.

This linguistic metaphor was quite common in the nineteenth century; among others, the famous German palaeontologist Heinrich Georg Bronn (1800–1862) also used it in his *Untersuchungen über die Entwicklungs-Gesetze der organischen Welt während der Bildungszeit unserer Erd-Oberfläche* (1858).

The imperfect and incomplete nature of the palaeontological record induced several palaeontologists to reflect on the epistemic notion of the fossil record and time. As a result, they came up with practices to overcome the incomplete nature of past records. Georges Cuvier identified the fossil record with monuments and coins to represent iconic frames of deep time; Bronn worked on the fossil record to obtain valuable biological patterns, which presented and made visible Earth's natural history.

Representing the Fossil Record as Monuments of Remote Time

The idea that geological time could be managed, if spatially treated, signified one important turning point in palaeontological practice. Baron Georges Cuvier (1769–1832) was one of the first to merge space with time. He intended to "burst the limit of time" (Cuvier, *Essay*, 1813: 3) to investigate its obscure parts. Cuvier developed comparative morphological investigations to reconstruct the inhabitants of the former epochs. These enquiries were based on the assumption that fossils were similar to "coins" or "monuments:" both objects witnessed an ancient past that needed to be deciphered.[10] Only by complex deciphering practices could this unknown dimension be read, understood, and contemplated.

Differing from classic antiquarians or earlier geognosts, however, Cuvier proposed a rigorous method to decrypt the unknown meaning hidden in the monuments of nature. As he wrote in his *Essay on the Theory of the Earth*:

> As an antiquary of a new order, I was obliged at once to learn the art of restoring these monuments of past revolutions to their original forms, and to discover their nature and relations; I had to collect and bring together in their original order, the fragments of which they consisted;

---

[10] More details about this analogy can be found in Sepkoski (2017) and Rudwick (2008).

> to reconstruct, as it were, the ancient beings to which these fragments belonged; to reproduce them with all their proportions and characters; and, lastly, to compare them with those which now live at the surface of the globe. (Cuvier, *Essay*, 1813: 1)

The practice that enabled the naturalist to restore the imperfect and incomplete fossilised monuments of the past was thus based upon comparative anatomy. By carrying out comparative analyses, the palaeontologist could indeed refer 'with perfect certainty' each fragmentarily fossilised bone to a specific species, genus, or family. At the same time, he could expand the biological hierarchy of living organisms, thus increasing its diversity. Once the morphological features of the petrified organisms had been recognised, the fossils could be displayed in lifelike poses (for instance, in museum halls or in compendia) to better convey vivid images of the past eras.

The "art of determining ... bones, or, in other words, of recognising a genus, and of distinguishing a species, by a single fragment of bone," (Cuvier, *Essay*, 1813: 4) made up Cuvier's entire work and characterised the palaeontological style of work until the so-called paleobiological revolution of the 1960s (Sepkoski and Ruse 2009; Sepkoski 2012; Tamborini 2015a). Cuvier's practice thus marked an essential turning point in the establishment of palaeontology as a biological science. Its long-lasting impact on the development of palaeontology can be illustrated by the fact that most of the palaeontological expeditions that characterised the nineteenth and twentieth centuries were based on the same morphological practice. First, palaeontologists came across imperfect and incomplete fragments of extinct organisms; second, they started restoring the original organisms by completing their visible features; third, they determined the relationships between these remains and the living species; fourth, they compared their findings and assembled the fragmentary pieces into a unitary form; finally, they positioned the extinct organisms in a lifelike pose to provide convincing images of the past. As a result, the recreated organisms—and no longer the individual fragmentary fossils—became monuments of Earth's unknown time (Fig. 6.1).

By following Cuvier's practice, time was thus assimilated to the spatial morphological features that were visible in the reconstructed extant organisms: Earth's deep time became visible in fascinating monuments of the past. As every monument, the reconstructed organisms symbolised remote events that were no longer accessible. By carefully comparing the

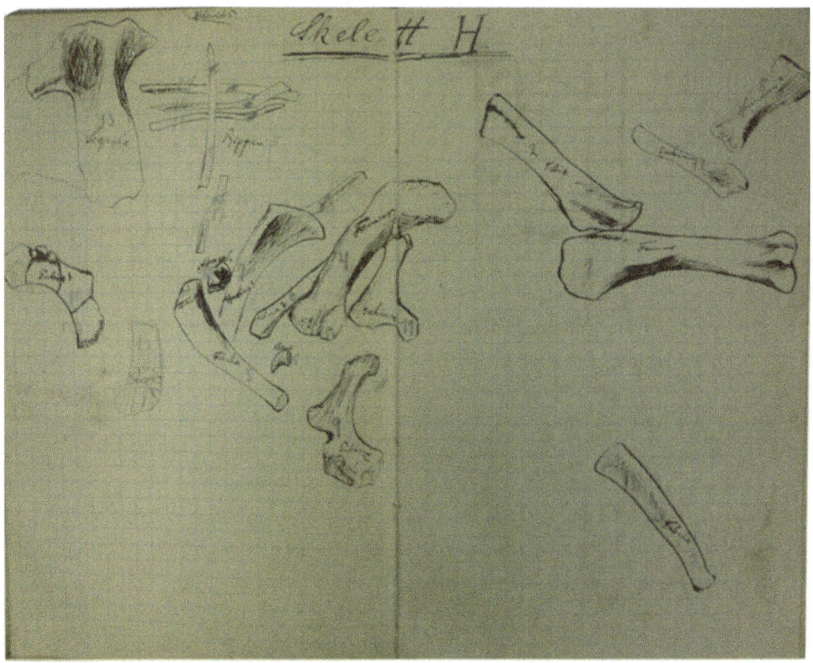

Fig. 6.1 One of the several palaeontological field sketches from the Tendaguru Expedition (1909–1913). The drawing clearly shows the fragmentary nature of the fossils found (MfN, HBSB, Pal. Mus, S II, Tendaguru-Expedition 8.2: 6)

fossilised bones of the extinct animals to reconstruct them, Cuvier took control over geological time: in short, he constructed a possible access to it, in effect re-presenting the record of the past. Within this practice, "re-presentation becomes a matter of presenting an initial *something* again and again; transforming, transposing, and translating the material/semiotic forms of that *something*; and serially disclosing and detailing what that initial, inchoate *something* was all along" (Lynch 2014: 324). Indeed, both the palaeontological field sketches and the illustrations of extinct organisms re-present an initial *something*, that is, the incomplete record of the past, eventually transforming and completing it through this process.

Furthermore, this practice made the represented temporal dimension accessible to a broader public. Reconstructed organisms, as, for instance, the mounted and exhibited skeleton of the big dinosaur *Brachiosaurus brancai* that is on display at the Berlin Museum of Natural History, could educate the general public, conveying a feeling of Earth's deep time and history (Fig. 6.2).

In fact, the *Brachiosaurus brancai* was explicitly exhibited according to Cuvier's ideal of science. In his inaugural address, German pathologist and member of the Tendaguru committee David Paul von Hansemann (1858–1920) asserted that, on seeing the bones, the expert recognised them as a work of art created by nature.[11] According to von Hansemann,

> the meaning of these findings goes well beyond its possible value for palae-ontology. In my opinion there is no better way to ethically educate masses than to show them works of art [as these dinosaurs' bones] in a very simple way. (von Hansemann, 1911: n.p.)

Each individual fossil, but especially the specific and artful arrangement of an assembly, thus became a marvellous icon of the deep past. Through its material presence, the spectator could access an even more unknown and intangible dimension: Earth's natural history. This dimension, however, was only partially visible through the comparative morphological method adopted by Cuvier. Hence, following Cuvier's practice, palaeontologists were obliged, as it were, to analyse this remote historical dimension. Which method, however, was the most appropriate to illustrate and read Earth's deep history?

The English palaeontologist and geologist John Phillips (1800–1874) raised this methodological point in 1837:

> The chronology adopted by geologists is liable to an inherent uncertainty or indefiniteness, quite different in its nature from the sources of error in ancient history. In the history of human affairs, the whole period which elapsed between the two epochs chosen as limit is known or supposed to be so; but the intervening occurrences cannot often be correctly placed in their true succession. In geology, on the contrary, the whole period included between the limits is, and perhaps must ever be, absolutely unknown; yet the succession of occurrences is, in general, clearly ascertained. (Phillips, *Treatise*, 1837: 291)

---

[11] On the historical, political, and social preconditions of this expedition and on its meaning for German science, see Tamborini (2016b).

Fig. 6.2 1937 display of the *Brachiosaurus brancai* in the Museum für Naturkunde Berlin (MfN, HBSB, Pal. Mus, B III/88)

With his observations, Phillips drew attention to a key argument: palaeontology only knew a merely chronological disposition of elements. Moreover, this structure was the result of complex practices that reconstructed the elements of past frames which were no longer available. However, what gave meaning to it, that is, the whole period that delimits and makes sense of this succession, is completely unknown. The key question was therefore as follows: How was it possible to proceed from the known to the unknown and backwards? What is actually more important: the mere, but known, chronological succession of events and organisms given in compendia, textbooks, or museum halls; or the unknown temporal period whose exhibition gives meaning to the known?

What was at stake was nothing less than the fundamental question of how to correctly narrate the history of nature. How should the fossil record, which represents the material concretisation of the unknown time, be read and used? Phillips and other prominent palaeontologists of the nineteenth century came up with visual practices apt for categorising unknown time. They understood that they had to develop techniques to narrate what happened in the unknown temporal dimension to overcome their fascination with time. Therefore, they developed practices that could adequately observe, manipulate, and represent unknown time. In doing so, they used epistemic tools such as tables, graphs, and diagrams to generate different data and access Earth's natural history.

TABLES, DIAGRAMS, GRAPHS, AND PATTERNS: PRESENTING THE FOSSIL RECORD IN THE FLOW OF UNKNOWN TIME

During the nineteenth century, the central question in geological sciences concerned the adequate method to grasp Earth's unknown history. Several palaeontologists addressed this problem, stressing the key notions of perception and of quantitative treatment of data. Perception is a fundamental activity, and the kind of answer that is given to the question of the modality of palaeontological observation determines what can effectively be done with the fossil record. Regarding this topic, Phillips had pointed out a clear dichotomy in deep time research: palaeontologists can either investigate the known succession of events or the entire unknown period in which this succession took place (Phillips, *Treatise*, 1837). But which approach was the more appropriate to grasp the immensity of geological time? Several palaeontologists were convinced that Earth's history could be grasped and understood only by quantitatively treating the fossil record. This practice identified possible numerical patterns in deep time, which in turn could be translated visually. As a result, the palaeontologist no longer acquired an assembly of more or less complex timeframes, but rather effectively developed a narration of Earth's natural history.[12]

Various palaeontologists developed practices based on a tabular classification of fossils to overcome the destructive nature of palaeontological time. These tables were not monographic studies that displayed organisms in a lifelike position (à la Cuvier). Neither were they descriptive reports on the species that had been found in particular rock layers nor data presented in a palaeontological or geological textbook. Quite on the contrary, they were lists of fossils found in various layers of the Earth. A typical example of these palaeontological tables can be seen in Fig. 6.3: In his *Paläozoologie: Entwurf einer systematischen Darstellung der Fauna der Vorwelt* (1846), the German palaeontologist Christian Gottfried Giebel (1820–1881) patiently calculated and tabulated all the fossilised species and genera that had ever been found at the time of his writing. For instance, he counted all the found species of the genus *Asaphus* (a genus of group of extinct marine arthropods) and arranged them according to the "place," that is, the "point in time," in which they had been found (first three columns). The last column indicates how many species have been unearthed in total (12 in this particular case).

The system of tables and the lists of dates were only materials required for further etiological elaborations. They generated evidence about what happened in the past: they offered, so to speak, a static natural history. To a certain extent, they are comparable to Cuvier's reconstructed organisms. As a result, they did not suffice to accomplish the most important and difficult task, which was, as Phillips had determined, to access unknown time to grasp the entirety of the geological period. Tables and data required yet another treatment, if one wanted to use them to narrate a natural history.

One of the first to understand the importance of quantitatively treating the fossil record was Heinrich Georg Bronn. He made exhaustive

---

[12] For a more detailed treatment of nineteenth-century quantitative palaeontology, see Tamborini (2015b and 2016a), Sepkoski (2013).

	Silurisches und devonisches Gebirge.	Kohlengebirge.	Kupferschiefergebirge.	Summa der Arten und Gattungen.
d) Polythalamia.				
1. *Orthoceratites.*	—	—	—	9
Orthoceras Breyn.	78	40	—	118
Lituites Montf.	8	—	—	8
Gomphoceras Murch.	3	—	—	3
Cyrtoceras Münst.	19	10	—	29
Gyroceras Meyer.	—	3	—	3
Actinoceras Bigsb.	1	—	—	1
Spirula Goldf.	11	—	—	11
Conoceras Bronn.	2	—	—	2
Phragmoceras Münst.	5	—	—	5
2. *Goniatitidae.*	—	—	—	3
Nautilus Lin.	5	36	—	41
Goniatites Hahn.	84	31	—	115
Clymenia Münst.	51	1	—	52
II. ARTHROZOA.				
4. ARTHROZOA.				
a) **Palaeadae.**				
1. *Trilobitidae.*	—	—	—	20
Trinucleus Murch.	8	—	—	8
Ogygia Brong.	2	—	—	2
Odontopleura Emmr.	3	—	—	3
Arges Goldf.	2	—	—	2
Brontes id.	4	—	—	4
Paradoxides Brong.	6	—	—	6
Olenus Burm.	8	—	—	8
Conocephalus Zenk.	2	—	—	2
Ellipsocephalus id.	1	—	—	1
Harpes Goldf.	1	—	—	1
Calymene Brong.	7	—	—	7
Homalonotus Kon.	5	—	—	5
Cyphaspis Burm.	1	—	—	1
Phacops Emmr.	20	—	—	20
Aeonia Burm.	3	—	—	3
Illaenus id.	3	—	—	3
Archegonus id.	4	—	—	4
Asaphus Brong.	12	—	—	12
Ampyx Dalm.	3	—	—	3
Phillipsia Kon.	—	6	—	6

Fig. 6.3 An example of tabular visualisation of the fossil record (Giebel, *Paläozoologie*, 1846: 76)

use of biochronological tables and developed a quantitative practice that enabled him to master the representation of the palaeontological remote past. First, Bronn collected and tabulated a great number of fossils in his *Index Palaeontologicus, oder Übersicht der bis jetzt bekannten fossilen Organismen* (1848–1849). This work was composed of an *Enumerator* (a list of all plant and animal fossilised species, chronologically and systematically ordered in tables) and a *Nomenclator* (an alphabetic list of the fossils record). In assembling these lists, Bronn approached the fossil record as a record of data that could be used to obtain valuable knowledge about the past: Earth's natural history could be narrated by reading the fossil record through its chronological setting, its systematic position, or its geographical dispersion. In fact, a combination of all these different perspectives constituted Bronn's meta-data in his tabular systems. Second, he decided to treat the tabulated data in a quantitative manner to calculate potential mathematical patterns. In the last volume of his *Handbuch der Geschichte der Natur* (1841–1849), Bronn finally practiced a statistical treatment of data, enabling him (and thus any other palaeontologist as well) to compare the number of plant and animal fossils with the living specimen (§ 7), count the total number of families (§ 8), and determine the ratio between families and species (§ 10). In other words, Bronn's quantitative treatment of data not only redefined what had to be understood as palaeontological data as such but also narrated the history of nature in chronological order. As a result, the palaeontologist also obtained mathematical patterns from the fossil record, which, in turn, assumed a biological meaning.

In addition, Bronn chose to represent these mathematical numbers spatially to visualise the past. For instance, he illustrated the history of the previously tabulated genus *Venus* (saltwater clams): its population steadily increased during all the geological epochs, reaching its peak only in the author's presence. The following diagram (Fig. 6.4) presents the appearance and disappearance of the classes of the animal and plant kingdoms. Graphs and charts became thus the means that allowed palaeontologists to master deep geological time and make it visible. These images presented "the working object[s]" (Daston 2014: 321) of palaeontology.

Hence, by tabulating the fossil record and reading it in a quantifying approach, Bronn generated visual patterns that presented Earth's natural history. His practice paved the way for ensuing palaeontological investigation into deep time. The immense palaeontological time could

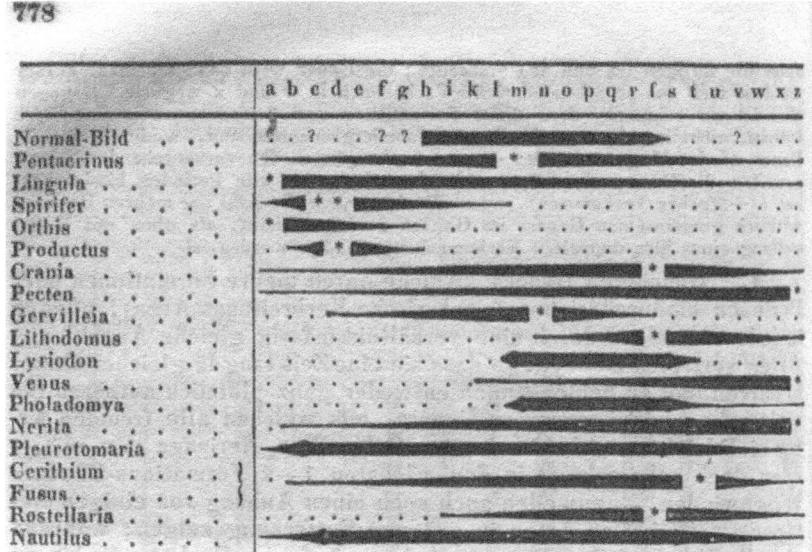

Fig. 6.4 Bronn's table of the development and the distribution of various genera (Bronn, *Index palaeontologicus*, 1849: 778)

be grasped only by presenting it as a series of visual and quantitative patterns.

John Phillips himself practiced this method. In *Life on Earth: Its Origin and Succession* (1860), Phillips used the same method as Bronn to narrate Earth's natural history. He acquired a sufficient amount of data by collecting as many fossils as he could, and on the basis of these data he compiled a "general table representing the numerical prevalence in time of each of the classes of Marine Invertebrata in the Lower Palæozoic Strata as at present known" (Phillips, *Life*, 1860: 78).

Successively, by using a table that gave to each class "a space proportioned to the number of species" (ibid., 79), Phillips visually showed what happened in palaeontological time: he made visible and narrated Earth's unknown history (Fig. 6.5).

This data-driven visual practice was not confined to European-continental palaeontology but was equally widely used by North American palaeontologists during the last part of the nineteenth and the entire twentieth century. Yale palaeontologist Henry Shaler Williams (1847–1918),

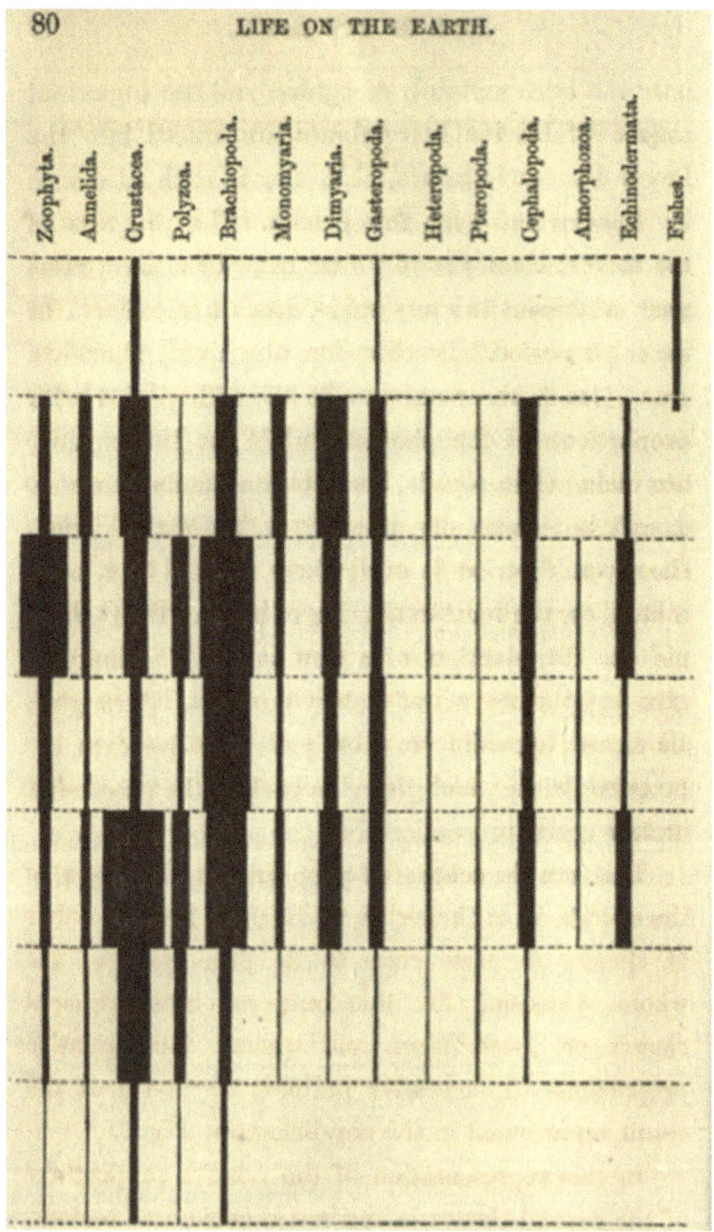

Fig. 6.5 Phillips' depiction of Earth's unknown history (Phillips, *Life*, 1860: 80)

	C.	O.	S.	D.	Cr.	T.	J.	K.	Ty.	Q.	R.
Tetracoralla ...81 genera	5	4	42	10	19	0	0	0	0	1	
Hexacoralla ...367 genera	0	4	17	11	7	22	72	75	81	78	
1. Favositidæ	0	3	14	8	3	0					
2. Poritidæ	0	1	2	0	2	1	2	5	8	3	
3. Madreporidæ	0					0			1	1	
4. Pocilloporidæ				2	1				1	1	
5. Eupsammidæ			1				2	2	6	7	
6. Fungidæ						6	13	9	9	19	
7. Astræidæ				1	1	13	46	44	33	28	
8. Styloporidæ							2	0	1	1	
9. Oculinidæ							6	2	4	4	
10. Dasmidæ									1	0	
11. Turbinolidæ						2	1	13	17	14	
Total Madreporaria ...448 genera	5	8	59	21	26	22	72	75	81	79	

Fig. 6.6 The table enumerates (regrouped in families) the first appearance of the order Madreporaria (marine animals) within a geological timeframe (Williams, *Geological Biology*, 1895: 84)

for instance, proceeded in exactly the same way.[13] He listed the kind and number of fossils that had been found and that dated from a particular era; based on these data, he then drew a graph to depict the rate of distribution of the genera in question (Fig. 6.6). Afterwards, the resulting table could be transformed into a graph that helped in pointing out the features of this particular historical pattern.

The resulting graph shows the growth of the marine animals of the order Madreporaria during the entire geological Phanerozoic eon, which began 542 million years ago. In addition, it illustrates the overall degree of development and differentiation of the complete order. Hence, through first an accurate tabulating of data and second a quantitative treatment, palaeontologists could do nothing less than present remote time. As a result, they could provide useful narrations, which finally helped them to escape from what Buffon had called the "dark abyss of time."

---

[13] Williams explicitly stated that "if we only note the numerical relation of these genera to the successive geological periods of time, the law above referred to becomes at once apparent" (Williams, *Geological Biology*, 1895: 84). This means that to come up with laws of development or distribution, the palaeontologist needs to put his data into numerical relations.

CONCLUSION

Since Cuvier's investigations, palaeontologists have adopted a variety of different strategies to visualise their data and come up with coherent descriptions of Earth's "deep time." The epistemic access to it has always been problematic. Indeed, as American geologist James Beerbower asserted, "yet as the chisel of a sculptor obliterates in succeeding blows its earlier marks, so these processes [Earth's natural processes] obliterate the traces of their own action, leaving the palaeontologist only the *smallest sample of the past on which to base his reconstruction*" (1960: 7). The natural processes inevitably destroy the traces of the past, and palaeontologists need to find a solution if they want to use the remaining data to decipher Earth's history. Quite early on, palaeontologists understood that they had to integrate the fragmentary nature of the fossil record with new data if they wanted to overcome the destructive effects of time. In fact, neither Cuvier's morphological analyses nor Bronn's quantitative practice did take the fossil record at face value—they did not literally 'read it.'[14] Rather, they supplied the loss of information generated by the dark abyss of time, generating new data that relied heavily on their practices and background knowledge. For instance, to mount and display the *Brachiosaurus brancai* in the exhibition hall of the Museum of Natural History at Berlin (see Fig. 6.2), palaeontologists actually 'generated' all the unpreserved records of this extinct organism: they came up with practices that allowed them to recreate the missing bones and put them together (Remes et al. 2011).

However, not every epistemic tool can equally access Earth's unknown history. According to the divergent scientific practices used to grasp the remote past, different spaces of empirical knowledge can be opened up. For instance, through a morphological reconstruction of the fossil record, the palaeontologist obtained only a series of individual temporal points of reference. Interpreted as monuments, these *represent* a quite definite span of time or even specific moments. For instance, the mounted skeleton of the dinosaur *Brachiosaurus brancai*

---

[14] Historian of science David Sepkoski defined the literal reading of the fossil record as a practice in which the fossil record "with all its notorious gaps and inconsistencies, was taken at face value as a reliable document. There never were, in other words, any missing pages or volumes; the discontinuities in the fossil record existed because the history of life is discontinuous" (Sepkoski 2012: 3).

in Berlin represents how life on planet looked 150 million years ago. As a monument, it represents this remote dimension. By tabulating and quantitatively treating these data, on the other hand, palaeontologists narrated complex temporal patterns of development, which *present* a different and more accurate image of this dimension. As a consequence, not *what* is given, that is, the fragmentary record of the past, but *how* it is *given* (Goodman 1978: 6) becomes the essential issue in understanding the features of the remote past.[15] By analysing *how* palaeontologists reorganised and analysed their data, I have pointed out two diverging visual cultures,[16] which in turn characterised the reasoning and the practices of numerous palaeontological sub-disciplines.

On the one hand, palaeontologists established an approach that was based on the more or less immediate translation of perception into representation: they reproduced what they perceived in rock layers. As Daston and Galison noticed, in this reproductive practice "the prefix *re-* is essential: images that strive for representation present again what already is" (2007: 382). Indeed, the champions of this method were palaeontologists such as Georges Cuvier or Karl Alfred von Zittel (1839–1904), who patiently presented again and again the morphological and physiological qualities of extinct animals to produce catalogues and compendia or to display them in lifelike poses. Cuvier, Zittel, and their followers meticulously pictured the morphological features of extinct animals and used their mimetic representations as genuine palaeontological data to grasp individual snapshots of geological time. They "purify, perfect, and smooth to get at being, at 'what is'" (Daston and Galison 2007: 382).

On the other hand, several palaeontologists established a visual culture that was centred on the quantitative elaboration of data. In doing so, they hoped to be able to go beyond what they could directly perceive in the fossil record. They ended up visualising broader patterns of temporal development. Graphs, tables, databases, charts, and statistical curves allowed palaeontologists to see patterns and processes that were not intuitively visible in their individual data. This second strategy

---

[15]About the relationship between Goodman's ways of worldmaking and unknown time, see Chap. 1 (Time in the Making) and Dorothee Xiaolong Hou and Sheldon H. Lu's article in this volume.

[16]On the notion of visual culture, see, for instance, Hentschel (2014), Heumann and Hüntelmann (2013). A detailed treatment of the visual language of nineteenth-century natural history can be found in Sepkoski and Tamborini, forthcoming.

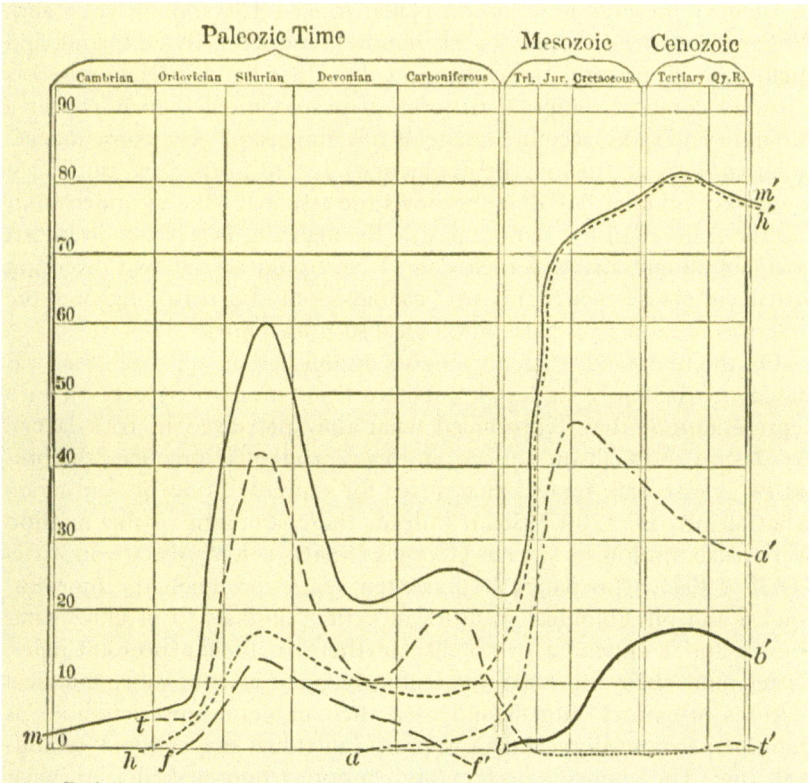

Fig. 6.7 The diagram shows the evolutionary curves of the families of Madreporaria: it is based upon the data of the previous figure (Williams, *Geological Biology*, 1895: 87)

proved to be an indispensable tool if one wanted to visualise complex historical patterns on the level of "deep time." Hence, the aim of these palaeontologists "was not simply to get the images right but also to manipulate the images" (Daston and Galison 2007: 382) through quantitative practices. As a result, these images did not represent again and again individual snapshots of time but sought to present the entire temporal dimension. Second, graphs and tables were no "longer necessarily focused on copying what already exists," such as the morphological features of the fossil record; instead, what they present became "part of a coming-into-existence" (ibid., 383). Through a graphical presentation of data, the evolutionary deep history of extinct genera such as the families *Madreporaria* could come into being. As a result, the images proper of this visual culture such as Fig. 6.7 were "examples of right depiction— but of objects that are being made, not found" (ibid., 391).

Hence, the general conclusions that emerge from my contribution are that (1) the unknown is the condition of possibility for palaeontological research. In palaeontology, the unknown is always coupled with time, because time destroys evidence of past epochs. (2) Not to lose their own way in the vast extent of time on a geological scale, palaeontologists tried to spatially grasp its immensity: our understanding of "deep time" is therefore inseparable from the techniques we employ to represent and present it by way of spatial models such as depictions or patterns in graphs and tables. Although we are not able to visualise time without a spatial dimension, (3) not every technique to measure, access, manipulate, and represent the unknown is adequate in the same manner. In fact, Cuvier and Phillips accessed, so to speak, two different and variously rich versions of the past. Hence, to understand the features of geological time, the key question must be the following: How has the palaeontological representation and presentation of data changed over time and under which social, political, economic, and epistemic conditions?

Acknowledgments This work was supported by the Federal Ministry of Education and Research, Germany, (BMBF) collaborative research project: *DiB—Dinosaurier in Berlin. Der Brachiosaurus brancai—eine politische, wissenschaftliche und populäre Ikone.*

BIBLIOGRAPHY

Beerbower, James R. 1960. *Search for the Past: An Introduction to Paleontology.* London: Prentice-Hall International.

Benton, Michael J. 1997. Models for the Diversification of Life. *Trends in Ecology & Evolution* 12: 490–495.

Blumenbach, Johann Friedrich. 1788. *Handbuch der Naturgeschichte.* Göttingen: Dieterich.

Bronn, Heinrich Georg. 1849. *Index palaeontologicus oder Übersicht der bis jetzt bekannten fossilen Organismen.* Stuttgart: Schweizerbart'sche Verlagsbuchhandlung.

————. 1858. *Untersuchungen über die Entwickelungs-Gesetze der organischen Welt während der Bildungs-Zeit unserer Erd-Oberfläche.* Stuttgart: Schweizerbart'sche Verlagsbuchhandlung.

Cardani, Michele, and Marco Tamborini. 2016. Data–Phenomena: Quid Juris? *Zeitschrift für philosophische Forschung* 70 (4): 527–548.

Cleland, Carol E. 2002. Methodological and Epistemic Differences between Historical Science and Experimental Science. *Philosophy of Science* 69 (3): 447–451.

Comte de Buffon, Georges-Louis Leclerc. 1778. Les époques de la nature. In *Histoire naturelle.* Paris: Imprimerie royale.

Cuvier, Georges. 1813. *Essay on the Theory of the Earth.* Edinburgh: William Blackwood.

Darwin, Charles. 1964. *On the Origin of Species.* Cambridge, MA: Harvard University Press.

Daston, Lorraine. 2014. Beyond Representation. In *Representation in Scientific Practice Revisited,* ed. Catelijne Coopmans, Janet Vertesi, Michael Lynch, and Steve Woolgar, 319–322. Cambridge, MA: MIT Press.

Daston, Lorraine, and Peter Galison. 2007. *Objectivity.* New York: Zone Books.

Efremov, I.A. 1940. Taphonomy: A New Branch of Paleontology. *Pan-American Geologist* 74: 81–93.

Engels, Eve-Marie, and Thomas F. Glick (eds.). 2008. *The Reception of Charles Darwin in Europe.* The Athlone Critical Traditions Series: The Reception of British and Irish Authors in Europe, vol. 1. London: Continuum.

Giebel, Christoph Gottfried Andreas. 1846. *Paläozoologie: Entwurf einer systematischen Darstellung der Fauna der Vorwelt.* Merseburg: Rulandt'sche Buchhandlung.

Gitelman, Lisa (ed.). 2013. *"Raw Data" Is an Oxymoron.* Cambridge, MA: MIT Press.

Goodman, Nelson. 1978. *Ways of Worldmaking.* Hassocks: Harvester Press.

Hansemann, David von. 1911. Geheimes Staatsarchiv Preußischer Kulturbesitz, Berlin.

Hartwig, Georg. 1863. *Die Unterwelt mit ihren Schätzen und Wundern: Eine Darst. für Gebildete aller Stände.* Wiesbaden: C.W. Kreidel's Verlag.

Hentschel, Klaus. 2014. *Visual Cultures in Science and Technology: A Comparative History.* Oxford: Oxford University Press.

Heumann, Ina, and Axel C. Hüntelmann. 2013. Einleitung: Bildtatsachen: Visuelle Praktiken der Wissenschaften. *Berichte zur Wissenschaftsgeschichte* 36 (4): 283–293.

Keller, Gerta. 2005. Impacts, Volcanism and Mass Extinction: Random Coincidence or Cause and Effect? *Australian Journal of Earth Sciences* 52: 725–757.

Kohn, David. 1985. *The Darwinian Heritage.* Princeton: Princeton University Press.

Lynch, Michael. 2014. Representation in Formation. In *Representation in Scientific Practice Revisited*, ed. Catelijne Coopmans, Janet Vertesi, Michael Lynch, and Steve Woolgar, 323–328. Cambridge, MA: MIT Press.

Massimi, Michela. 2011. From Data to Phenomena: A Kantian Stance. *Synthese* 182 (1): 101–116.

Matthew, William Diller. 1915. *Dinosaurs with Special Reference to the American Museum Collections*. New York: American Museum of Natural History.

Müller-Wille, Staffan, and Isabelle Charmantier. 2012. Natural History and Information Overload: The Case of Linnaeus. *Studies in History and Philosophy of Science Part C: Studies in History and Philosophy of Biological and Biomedical Sciences* 43 (1): 4–15.

Phillips, John. 1837. *A Treatise on Geology: Forming the Article under that Head in the Seventh Edition of the Enyclopaedia Britannica*. Edinburgh: Adam and Charles Black.

———. 1860. *Life on the Earth: Its Origin and Succession*. Cambridge: Macmillan.

Remes, Kristian, David M. Unwin, Nicole Klein, Wolf-Dieter Heinrich, and Oliver Hampe. 2011. Skeletal Reconstruction of *Brachiosaurus brancai* in the Museum für Naturkunde, Berlin: Summarizing 70 Years of Saurapod Research. In *Biology of the Sauropod Dinosaurs: Understanding the Life of Giants*, ed. Nicole Klein, Kristian Remes, Carole T. Gee, and P. Martin Sander, 305–316. Bloomington: Indiana University Press.

Rheinberger, Hans-Jörg. 2011. Infra-Experimentality: From Traces to Data, from Data to Patterning Facts. *History of Science* 49 (3): 337–348.

Rossi, Paolo. 1984. *The Dark Abyss of Time: The History of the Earth & the History of Nations from Hooke to Vico*, trans. Lydia G. Cochrane. Chicago: The University of Chicago Press.

Rudwick, Martin J.S. 1972. *The Meaning of Fossils: Episodes in the History of Palaeontology*. London: Macdonald.

———. 1985. *Scenes From Deep Time: Early Pictorial Representations of the Prehistoric World*. Chicago: The University of Chicago Press.

———. 2005. *Bursting the Limits of Time: The Reconstruction of Geohistory in the Age of Revolution*. Chicago: The University of Chicago Press.

———. 2008. *Worlds Before Adam: The Reconstruction of Geohistory in the Age of Reform*. Chicago: The University of Chicago Press.

Scrope, George Julius, and Poulett. 1827. *Memoir on the Geology of Central France: Including the Volcanic Formations of Auvergne, the Velay, and the Vivarais*. London: Longman, Rees, Orme, Brown, and Green.

Sepkoski, David. 2012. *Rereading the Fossil Record: The Growth of Paleobiology as an Evolutionary Discipline*. Chicago: The University of Chicago Press.

———. 2013. Towards 'A Natural History of Data': Evolving Practices and Epistemologies of Data in Paleontology, 1800–2000. *Journal of the History of Biology* 46 (3): 401–444.

————. 2017. The Earth as Archive. In *Archiving Sciences: Pasts, Presents, Futures*, ed. Lorraine Daston. Chicago: The University of Chicago Press.

Sepkoski, David, and Michael Ruse (eds.). 2009. *The Paleobiological Revolution: Essays on the Growth of Modern Paleontology*. Chicago: The University of Chicago Press.

Sepkoski, David, and Marco Tamborini. 'An Image of Science': Cameralism, Statistics, and the Visual Language of Natural History in the Nineteenth Century. *Historical Studies in the Natural Sciences* (forthcoming).

Tamborini, Marco. 2015a. Die Wurzeln der ideographischen Paläontologie: Karl Alfred von Zittels Praxis und sein Begriff des Fossils. *NTM Zeitschrift für Geschichte der Wissenschaften, Technik und Medizin* 23: 117–142.

————. 2015b. Paleontology and Darwin's Theory of Evolution: The Subversive Role of Statistics at the End of the 19th Century. *Journal of the History of Biology* 48 (4): 575–612.

————. 2016a. 'Nur eine bloße Anhäufung von gewonnenen Thatsachen': Erste Versuche eine quantitative Paläontologie zu begründen. *Mitteilungen der Österreichischen Gesellschaft für Wissenschaftsgeschichte* 32: 192–205.

————. 2016b. 'If the Americans Can Do It, So Can We': How Dinosaur Bones Shaped German Paleontology. *History of Science* 54 (3): 225–256.

Turner, Derek. 2005. Local Underdetermination in Historical Science. *Philosophy of Science* 72 (1): 209–230.

Williams, Henry Shaler. 1895. *Geological Biology*. New York: Henry Holt and Company.

Woodward, Samuel Peckworth. 1856. *A Manual of the Mollusca; Or, A Rudimentary Treatise of Recent and Fossil Shells*. London: John Weale.

PART II

Unknown Presents

Introducing Time to Ethnographic Displays: Narrative Strategies of Revealing the Unknown in German Ethnological Museums

Katja Wehde

INTRODUCING TIME

In the article "Of Other Spaces: Utopias and Heterotopias," Foucault conceptualises the modern museum as a "place of all times that is itself outside of time and inaccessible to its ravages" (1986: 26). On the one hand, this observation is no longer applicable. The postmodern museum does not stand unaffected by the passage of time, but it is increasingly confronted with public demands, participatory challenges, and academic criticism. Both the public role of the institution and the meaning of its exhibits are constantly changing.

On the other hand, Foucault's statement still holds true. Museums, and particularly ethnographic museums, display objects from all times while positioning themselves outside of time. They exhibit the history of their artefacts without revealing their own role in that same history.

K. Wehde (✉)
International Graduate Centre for the Study of Culture, Justus-Liebig-University, Giessen, Germany

Focusing on the portrayal of seemingly pure and original contexts, curators in ethnographic museums rarely make visible how the exhibited objects have arrived in the museum. In many institutions, the time of collecting art and artefacts hence remains unknown. The resulting time lapse between ethnological museums and their objects further manifests itself in what Johannes Fabian has called the denial of coevalness of the Other (1983: 34). In his foundational essay on "Time and the Other," Fabian argues that nineteenth-century ethnographic accounts represented Europe's Others in an undefined pre-modern time whereas ethnographers situated themselves within a specifically Western modernity. This critique has long been translated to the ethnographic gallery. In her work on *Translating Others*, Kate Sturge shows that exhibitions may represent non-European people and artefacts in an indeterminate and static past if they use unspecific time frames as well as the "Ethnographic Present" to explain cultural practices and products (2006: 433). Both these mechanisms of locating Others in an unspecified past and of excluding collection history from museum narratives have led to the construction of an unknown time in the ethnographic museum.

However, museums are not in fact impervious to the effects of time, which is why the self-representation of the ethnographic museum as an omniscient narrator has become an important point of criticism. As a result of academic and public demands for an institutional reformation, many museum professionals are increasingly willing to take the challenge to reveal current issues of collecting, preserving, and exhibiting to the public. In light of the success of participatory exhibition designs and museum documentaries, the institution has grown aware of the benefits of allowing visitors to look behind its scenes. This process can be related to the turn to the "post-museum" where visitor experiences and imagination play as big a role as expert knowledge (Hooper-Greenhill 2000: 142). In the context of this democratic opening of the museum as institution, ethnographic museums have undergone considerable changes including the opening of new galleries that feature representations of Europe in the now-renamed World Culture Museums. Additionally, some museums have recently launched projects and exhibitions that refer to their own institutional history or problematise the representation of colonial artefacts. For example, the Horniman Museum's Centenary Gallery in London "tells the story of the Horniman's collections and how they were viewed by different collectors over the last 100 years" ("Centenary Gallery," n.d.). Equally, the Rautenstrauch-Joest-Museum in Cologne has opened a permanent exhibition on ethnographic museum practices and appropriations ("Die Welt in der Vitrine: Museum," n.d.).

The ethnographic museum's grandstanding as an entity disassociated from the objects on display is hence no longer self-explanatory. Unknown time in the form of collection history is slowly being introduced to the ethnographic museum. This process is particularly interesting because it has the potential to renegotiate the long-established binary opposition between 'us' and 'them.' Portraying the role of European explorers and ethnographers in selecting and interpreting artefacts can thus be regarded as a measure to reframe the museum as a place of active meaning-making. Although Donald Preziosi argues that "the museum could well persuade us to believe that there was a real history out there independent of our historiographies, our museographies" (1996: 106), the introduction of collection history could lead to a reconsideration of the institution's interpretational sovereignty.

However, the representation of the museum's contested history could equally evoke critical questions about the ownership of non-European artefacts. By exposing ethnography's function as a form of "culture collecting," the presentation of collection history may reveal "the ways that diverse experiences and facts are selected, gathered, detached from their original temporal occasions, and given enduring value in a new arrangement" (Clifford 1988: 231). These notions of selecting facts and meanings stand in contrast to the museum's current public role as a reliable mediator of knowledge and might hence call into question the integrity and overall raison d'être of the institution.

How do museums that address contested practices of explorers and ethnographers in the nineteenth century make known this unknown time without affecting their public perception in the present? In the following, two different means of introducing collection history are analysed in more detail. The analysis explains narrative strategies of telling an unknown time and investigates the ways in which a distance between the portrayed contested history and contemporary museum practices is established. Furthermore, it reflects on a potential recreation of the fascination with unknown time, which may result from confronting visitors with exotifying or romanticised ethnographic material. How can narratives of collection history prevent a renewed exoticisation of the Other or the evocation of a nostalgia for the time of exploration and discovery?

The meaning of unknown time in this analysis is twofold. On the one hand, details about collecting practices are in fact often unknown because they have not been documented (Gerstenblith 2011: 451). On the other hand, ethnographic museums have avoided displaying critical aspects of their collection history to prevent falling into public disrepute. How do the case studies explored in this analysis fill these gaps in knowing and showing?

CASE STUDIES: *TOTEM'S SOUND* AND *FOREIGN EXCHANGE*

The first case study is a temporary exhibition of the Weltkulturen Museum in Frankfurt entitled *Foreign Exchange, or the Stories You Wouldn't Tell a Stranger*. The exhibition was presented from January 2014 to January 2015. The founding director of the institution, Bernhard Hagen, had worked as a doctor in the plantations of Southeast Asia in the nineteenth century. His ethnographic photographs, stored in the archives of the museum, raised questions about his real interests in working with the indigenous inhabitants on site (Deliss 2014: 11–12). The exhibition contrasts ethnographic artefacts and photographs from the collection with contemporary pieces of art. It addresses the circumstances under which ethnography was conducted and documents aspects of colonialism, subordination, and resistance.

The second case is the computer game *Totem's Sound*, which was developed by the Ethnologisches Museum Berlin-Dahlem in collaboration with the Humboldt Lab, a project dedicated to testing new exhibition designs for the collection's future in the Humboldt Forum at the heart of Berlin. The game was part of the project "Reisebericht" in the laboratory exhibition *Probebühne 4* (September 2014–February 2015) and is available for download online (*Totem's Sound* 2012). *Totem's Sound* follows the quest of the collector Johan Adrian Jacobsen, who was commissioned by the Ethnologisches Museum to gather artefacts from North America in the nineteenth century. In the course of the game, Jacobsen talks to indigenous villagers, experiences adventures, and passes tests to retrieve as many objects as possible in one day. The game's narrative consists of five chapters that are all geared towards the acquisition of a particular artefact from the villagers.

An important difference between the two case studies obviously exists with regard to their contexts. Although Whereas *Foreign Exchange* was a complete exhibition, *Totem's Sound* was only part of a whole. In addition to the computer game, the section "Reisebericht" in *Probebühne 4* included an augmented reality presentation as well as a puppetry film entitled *Der von einem Stern zum anderen springt*. Both projects equally drew on Johan Adrian Jacobsen's travelogue. *Totem's Sound* can hence be regarded as a single exhibit and may therefore not seem comparable to a complete exhibition. However, the game is still available for download online although the temporary exhibition has ended. Potential players are thus not able to experience the game in its original context. Furthermore, the computer game may have been complemented by the augmented reality presentation and the puppetry film, but the multifaceted depiction of Jacobsen's voyage can be regarded as a reflection on the museum's collection history in its own right. For the purpose of comparing different narrative strategies of introducing unknown time, it thus makes more sense to analyse the game in detail instead of seeing it only as a minor part of the bigger temporary exhibition.

Another difference exists with regard to the explicitness of the institutional and disciplinary criticism that is articulated within the two examples. As gold extra, the game's production team, states in the press release for *Probebühne 4*, *Totem's Sound* is designed as a compact and narration-laden game that only partly revolves around historical details (quoted in König et al. 2014). As the game's narrative is merely inspired by the travelogue of Johan Adrian Jacobsen, the fictional adventure story only implicitly criticises the practices of the museum and its commissioned travellers. In contrast, *Foreign Exchange* facilitates a view from inside the museum by using documentary and artistic techniques. The exhibition hence directs criticism at itself and its own history more openly. Although the game draws upon real historical events and turns them into a fictional plot, the exhibition portrays historic documents and artefacts and turns them into an interpretative but nonfictional visual narrative. This display has some implications for the ways in which the different case studies may spark a fascination with the unknown time of collection history. The design of *Totem's Sound* as a fictional adventure game may evoke a somewhat positive fascination with notions of exploration and discovery, but the nonfictional account of *Foreign Exchange* constantly reminds the museum visitors of the real exploitation of non-European Others and thereby counters such reactions more immediately. Although *Totem's Sound* equally applies distancing devices that problematise early ethnographic practices, the instant visual impression of the point-and-click adventure creates an atmosphere of fascination (see Fig. 7.1).

Fig. 7.1 Screenshot of *Totem's Sound*, introduction (Gold Extra 2014, http://www.totems-sound.com/, last accessed January 6, 2017). In the framework of: Humboldt Lab Dahlem, Probebühne 4, Kulturstiftung des Bundes, Stiftung Preußischer Kulturbesitz, Berlin

Narrative Strategies of Introducing Collection History to the Ethnographic Museum

Having explained the differences between *Totem's Sound* and *Foreign Exchange*, I now discuss different narrative strategies applied in both case studies to introduce the unknown time of collection history. The analysis does not aim at delivering an extensive survey of all the strategies that are employed. In reference to the introductory questions, the focus is on those patterns and methods that are employed to bridge the gaps in knowledge and documentation as well as to disassociate current museum practices from those of the past.

MULTI-PERSPECTIVITY AND SUBJECTIVITY

As Mieke Bal has reasoned, narratives in exhibitions are based upon an "implied 'focalizer'" (1992: 561). In *Foreign Exchange*, this is the potential visitor who walks through the gallery space and thereby creates his or her own story. Although *Totem's Sound* cannot be experienced physically, the metaphor still applies because playing a computer game involves a virtual journey through the game's chapters. Andrew Burn has argued with respect to video game analysis that the storyworld of a game is most often perceived from the avatar's perspective, rendering it an internal focalisation (2014: 243). In contrast, exhibitions often work with different kinds of focalisation. Visitors may be guided by an internal focaliser who tells his or her own story with respect to the theme of the exhibition, but they might equally receive contextual and comparative information in the sense of a zero focalisation (Buschmann 2010: 153ff.).

In *Foreign Exchange*, we are presented with a predominantly external focalisation as we perceive the individual exhibits from an outside perspective. The narrative mainly works through the juxtaposition of artefacts from the collection with contemporary art and includes only a few introductory or explanatory gallery texts. The exhibition provides an aesthetic experience rather than an explicit interpretation of the imperialist world view of ethnologists and collectors at the time. By way of this external focalisation, the visitor is not only denied a comprehensive insight into the historical background but also an analytical interpretation of the depicted material. This denial of an explicit institutional commentary has recently been problematised by Michael Kraus, who has argued that a lack of academic analysis may expose the material to the visitors' own projections (2015: 235). Although museum education theories invite audiences to construct their own stories from exhibitions (Hooper-Greenhill 2000; Hein 2002), such concerns seem legitimate especially with regard to sensitive and political representations. As Annabel Fraser and Hannah Coulson ask: "If we invite visitors to draw their own conclusions, how comfortable are we about leaving room for them to get it 'wrong'...?" (2012: 228) This question is especially important with regard to the introduction of collection history into ethnographic displays. Does an exposure to romanticised and exotifying ethnographic photographs spark a positive fascination with the unknown time of "culture collecting"? How can artists and curators prevent the portrayal of ethnographic material from eliciting an "imagined nostalgia" (Appadurai 1996: 77) for a past that the visitors never experienced?

In the exhibition, this problem is solved by the contrast that is constructed through the comparison of ethnographic material with contemporary works of art. The artestic reflections speak to the artefacts and photographs, distort them, exaggerate their effects, or simply showcase the ideology behind them. As visual representations they may even be more immediate than a textual analysis or an in-depth contextualisation. The perception of ethnographic objects and photographs is mediated by contemporary artistic counter-impressions that prevent a renewed exoticisation. Furthermore, as various artists are featured in the exhibition, multiple perspectives on the ethnographic material are portrayed.

The exhibition hence finds a solution to the problem that the material presents. An equal and non-Eurocentric academic analysis would necessitate the description of historical details that either have not been sufficiently documented or which can only be reconstructed by referring to Western ethnographic writing. The voices of the people depicted in the ethnographic photographs could and would not be represented. Moreover, if the exhibits were presented along with in-depth analyses explaining the reasons for exploitation and control, this could contribute to a relativisation of the power imbalance between nineteenth-century ethnographers and their research subjects. In contrast, a multi-perspectival visual narrative can offer more immediate critical views without reinforcing the museum's interpretational control.

For example, a slide show developed from a group discussion between artists and ethnologists presents ethnographic photography from the museum's archive as well as contemporary photographs related to the topics of otherness, the ethnographic gaze, and power. The walls surrounding the projection of the slide show are additionally covered with quotes from the group discussion (Fig. 7.2). They address the historical background and circumstances of ethnography and collecting, and mirror the issues of scientific objectification, exoticisation, and resistance raised by means of the contemporary photographs. This strategy of granting visitors a more contextualised and commented encounter of the material is again marked by the multi-perspectivity of the quotes, which denies an impression of facticity and universality. The gallery offers visitors a context and allows an understanding of the processes of nineteenth-century collecting without laying claim to a single truth.

In *Totem's Sound*, the story unfolds from the perspective of Johan Adrian Jacobsen. By making viewers perceive and even interact with the fictional world through his eyes only, the computer game introduces

Fig. 7.2   Exhibition view of study room with quotes, Weltkulturen Museum Frankfurt, Jan 2014–Feb 2015 (Photograph courtesy of Wolfgang Günzel 2013)

unknown time in a subjective way. Similar to the presentation of multiple perspectives in *Foreign Exchange*, this approach avoids an omniscient positioning of both the museum and the visitors. Furthermore, this narrative strategy relates to the actual circumstances of introducing collection history. Jacobsen's account in his travelogue is indeed the only perspective that is documented. In reality and in the game alike, knowledge is bound to his writings, although these may be at odds with the truth. However, historians and visitors are able to construct their own interpretations of these documented experiences by adopting Jacobsen's perspective or reflecting upon the historical situation. Thereby, just like the travelogue, the game blurs fiction and reality, mirroring the ethnological problem of recapitulating the activities of former collectors and ethnologists.

Yet, the internal focalisation is not unproblematic. Johan Adrian Jacobsen is not only the main character but also the only avatar in the game, which elevates the agency of the white explorer over that of the indigenous characters featured in the story. Thereby, the game players are presented with a narrative world that renders Jacobsen as an adventurer whereas the experience of the Other might paint a different picture (as, for instance, of Jacobsen as thief or liar). Although the game portrays Jacobsen's obsession with artefacts as well as his nonempathic engagement with the indigenous inhabitants, these aspects appear only subtly and often take a backseat to the romantic quest with which the game players (in the form of the avatar Jacobsen) identify. The adventurous motif of the game's narrative may hence spark a positive fascination not only with the unknown time of collecting artefacts but also with the Other as an entity. Casting Jacobsen as an amateur adventurer by way of the game design (see Fig. 7.1) and the plot may further trivialise the ways in which museums have retrieved some of their objects. If the story in the computer game is largely fictional, why does the game not provide any other avatar such as a villager or a tribal woman?

IRONY

The use of irony stands out as a narrative strategy in both the game and the exhibition. *Totem's Sound* uses irony at various points, for instance, when Jacobsen obtains an object at the end of each chapter. The artefact is then presented on screen in the style of an advertisement or a home shopping item (Fig. 7.3). Through this ironical depiction of his acquisitions, the European attraction to rare and special artefacts is ridiculed by means of a visual reference to contemporary consumption practices. As in the case of the adventurous portrayal of unknown time, the question remains whether such a depiction raises awareness for or trivialises the incessant quest for exotic artefacts during the nineteenth century. Either way, as the irony is directed at the museum and its collectors, it offers an alternative perspective on the travelogue and Jacobsen's journey while at the same time marking this perspective as exaggerated. This presentation negotiates the various ways in which history could have unfolded and creates a distance between institutional practices now and then.

Irony is employed more subtly in other parts of the game. In Chap. 1, Jacobsen asks an indigenous woodcarver for an artefact that he would like to purchase from him. The man tells him to pay for the object

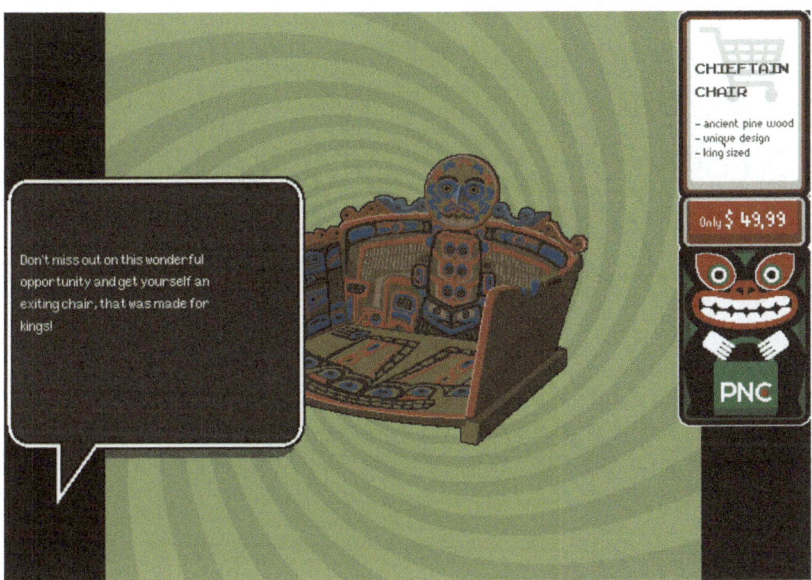

Fig. 7.3 Screenshot of *Totem's Sound*, Chap. 1 (Gold Extra 2014, http://www.totems-sound.com/, last accessed January 6, 2017)

with a colourful blanket that "some of the Old Hamatsa living in the forest" still make. The detail that "Hamatsa" is the name of a North American secret society said to practise cannibalism remains unknown to both Jacobsen and the game players. Although the phrase "some of the Old Hamatsa" hints at the fact that "Hamatsa" is the name for an entire community, Jacobsen refers to a member of the community as "Mr. Hamatsa" (Fig. 7.4). Thereby, Jacobsen translates the demonym "Hamatsa" into the family name "Hamatsa." As this episode points to issues of generalisation and individuality, it can be interpreted as an ironic play on the ethnographic practice of labelling and generalisation that Clifford Geertz has described by stating that ethnographers do not study villages, but *in* villages (1973: 22). The scene also portrays the misunderstandings, confusion, and negotiation of meaning that underlie processes of encounter and serves as an example of sources of error that present themselves in ethnographic fieldwork.

Jacobsen must subsequently go to the forest and succumb to the biting ritual to retrieve the blanket with which he can pay the woodcarver for an

(a)

those. Not for stupid coins or
tobacco. Maybe you can get some of the
old Hamatsa living in the forest.

(b)

Jacobsen: What? Why would this Mister
Hamatsa want to bite me?

Fig. 7.4   a *(Top)*, b *(Bottom)*. Screenshots of *Totem's Sound*, Chap. 1 (Gold Extra 2014, http://www.totems-sound.com/, last accessed January 6, 2017)

indigenous artefact. The Hamatsa he encounters laments that "only few people come to the forest to get bitten nowadays," upon which Jacobsen expresses his "deepest sympathies." The use of irony is more problematic here than in the first case as its reference point is the Hamatsa community instead of Jacobsen's behaviour or Europe's obsession with the exotic. By highlighting the abnormality of the biting ritual, the binary opposition between 'us' and 'them' is reestablished. Jacobsen's slightly condescending attitude towards the Hamatsa fails to be emphasised as such. In the end, the joke is on the Hamatsa. In this incident, the game falls into the trap that *Foreign Exchange* seeks to avoid when introducing unknown time from an either uncommented or multiperspectival view: it reinforces exoticisation in the process of addressing collection history.

While irony in *Totem's Sound* mainly works through the dialogues and texts, irony in *Foreign Exchange* is predominantly introduced by means of visual narratives. The aforementioned slide show involves ethnographic photographs from the museum's archive that Paul Schebesta and Martin Gusinde (both missionaries and ethnographers) took during their own field research. The images depict the ethnographers standing next to natives and measuring them by comparing their own height to the bodies of their research subjects. The slide show makes transparent the problematic dimension of ethnographic photography as an objectifying practice. At the same time, the exhibit prevents a renewed stigmatisation of the Other as well as a nostalgic fascination with the represented time; this is achieved at least to some extent with irony. Mixed into the slide show of ethnographic photography are contemporary photographs of artists reflecting upon issues of colonialism, the gaze, and resistance.

For example, Pushpamala N and Clare Arni's image *Toda* (http://artsearch.nga.gov.au/Detail.cfm?IRN=172053, last accessed March 17, 2017) ironically addresses ethnographic photography by staging a Toda woman in front of a chessboard background together with a measuring device. This representation is reminiscent of typical anthropometric studies of the nineteenth century, which are equally part of the slide show. In the photograph, "Pushpamala … enacts a performative mimesis through the lens of anthropology and confronts the viewer through the double gaze of a Toda woman and a contemporary artist" (Kaur and Dave-Mukherji 2014: 10). Juxtaposed with the colonial photographs in the projection, the image critically questions ethnographic practices and enables the alleged Other to take agency of her own depiction. The Other talks back and imitates the colonial depictions that appear in the slide show.

This ironic appropriation of ethnographic photography not only displays a critical self-reflection of the museum's history but manages to narrate this unknown time without re-essentialising and victimising the Other.

This empowerment of the Other is mirrored by one of the quotes from the conversations documented on the surrounding walls:

> In many travelogues one can read that the people were scared of being photographed or measured. Women were shaking, some jumped into the river. But there were different reactions as well: The native men began imitating the ethnologists. They started measuring the mules of the scientists, measuring their penises and reporting to the ethnographers. They were making fun of the process. (Michael Kraus, *Foreign Exchange*, Weltkulturen Museum Frankfurt, Jan 2014–Feb 2015 [my translation])

This account not only emphasises subtle strategies of resistance, which are often ignored in the depiction of colonial histories, but it also represents an ironic counterattack and thereby bridges the gap between the visual strategies in the contemporary images and the contextual historical information provided by the quotes on the surrounding walls. The recovering of unknown time thus unfolds in a way that does not entail master narratives or truth claims but which foregrounds the complexity of the history of collection and colonialism.

Another ironic exhibit within *Foreign Exchange* comes from Benedikte Bjerre, who arranged typical photographs of ethnographic artefacts next to Google image search results of the same objects. Ethnographic objects were traditionally photographed for the museum's archive using a special lighting and background colour. In a discussion of the work in the exhibition catalogue, Clémentine Deliss explains this with the desire of ethnologists to compensate a missing presence (Deliss and Mutumba 2014: 147). In the same conversation, Renée Mussai refers to the exhibited photographs as remystifications by means of lighting and colour: "We exotify [the objects] and ascribe them to a new place. This photographic style marks the attempt to call to mind the tribal aspects of an exotic environment" (ibid. [my translation]). Benedikte Bjerre's work critically confronts this practice in *Foreign Exchange*. Using reverse image search on Google by scanning a traditional image of an artefact and searching for comparable depictions, Bjerre displays the original photograph next to similar-looking Google images. Her work plays with ideas of association, representation, and exoticisation. The photograph of a bow holder

(a)   **(b)**

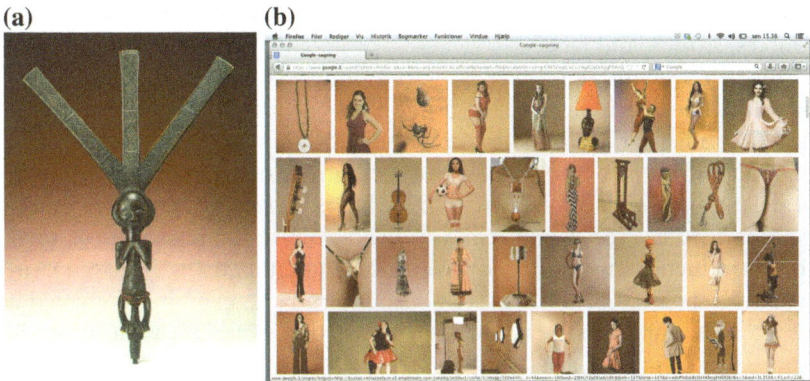

Fig. 7.5 a (*left*) Bow Holder. Acquisition of Guillaume Dehondt, 1942. Luba, Democratic Republic of Congo, Central Africa. Collection Weltkulturen Museum (Photograph: Maria Obermaier, 1989) **b** (*right*) Benedikte Bjerre 2013, *Google Image Search of 'Bow Holder'* (Photograph courtesy of the artist)

from the Congo, positioned in front of a dark red background, generates images of women in short dresses or provocative clothing (Fig. 7.5).

In the exhibition, the original photograph of the object in front of a red background does not become an object of exotic fascination but is exposed as a representative of an exotifying method. Fascination with unknown time is then only evoked in the sense of wondering to what lengths museums went to evoke the feeling of distance and difference. Constructing an almost innocent comparison between the resulting images by simply arranging them next to each other, the artist does not draw her own conclusions or preempt a statement on the nature of the similarities. Her work could be read simply as an ironic comment on such photographs, but it could also be interpreted as serious criticism of the legacy and implications of ethnographic practices. She thus introduces the unknown time of documentation and visual practices surrounding collecting history from a distance by giving little in-depth information and by opening up an ironic reflection on the matter.

AMBIGUITY

As already mentioned, information on collecting procedures, on the owners of the artefacts on display, and on the processes of acquisition are rarely documented sufficiently. Often, collecting time is indeed unknown. How do exhibition designs deal with these unorganised sources, stories, and facts?

In their recently published book on vagueness in violence, myth, and morality, Bernhard Giesen et al. argue that vagueness is not only a result of an abundance of information, representations, and images but can also be regarded as a method to "avert the danger of a violent production of definiteness" (2014: 13 [my translation]). In *Foreign Exchange*, ambiguity is equally used as a method to problematise the process of knowledge production and the knowledge the museum 'owns' today. David Weber-Krebs, one of the artists in residence for the exhibition, created a new textual database and a sound piece for the exhibition with his work *Immersion* (Fig. 7.6).

The artist created an inventory of the museum's collection by spontaneously assigning names to the artefacts. He explains his work as follows:

Fig. 7.6 David Weber-Krebs 2014, *Immersion* (Courtesy of the artist). Exhibition view, Weltkulturen Museum Frankfurt, Jan 2014–Feb 2015 (Photograph courtesy of Wolfgang Günzel 2013)

This work is a new form of creating inventories or the attempt to grasp the whole world. I have developed my own language, my own research language, my history. It mirrors the processes of systematisation applied originally to the collection by the museum. (quoted in Deliss and Mutumba 2014: 147 [my translation])

A part of the work consists of text lines on the gallery walls. The text is made up of all the names Weber-Krebs has assigned to the objects in the collection. Upon first entering the exhibition room, visitors merely perceive a grey pattern on the wall. As the writing appears extremely small in contrast to the large gallery space, the text lines may at first appear as wallpaper. Getting closer, however, does not reveal the entire story either. The text reads: "A hat for priests of the Amhara region in Ethiopia. Sandals of leather. A hat from the Philippines. It was collected in 1883" (ibid. [my translation]). It mainly features materials and regions of collecting, but collection dates and names of collectors are rarely available. The work thus points to a lack of information instead of revealing something about the labelled artefacts. The neutral style of the seemingly never-ending description and the grey visualisation on the walls make visible how much the museum owns and how little it knows. When taking a few steps back again from the wall, the lines blur and raise the questions: Is there anything more to the objects in the collection than the pattern created by listing them? Do these objects exceed mere representations of the obsession with foreign objects in the nineteenth century? Weber-Krebs's work does not provide an answer. Instead, the vagueness of both the visual representation and its textual content illustrate the uncertainty the ethnological museum as an institution is facing with regard to its practices, its archives, and its history. This work thus confronts the unknown dimension of history by visualising and materialising it.

Totem's Sound applies ambiguity both as a fact and as a method. In the aforementioned encounter with the Hamatsa, Jacobsen does not learn the reason for the biting ritual: this points to the indefinite as a fact. Ethnologists do not know much about the rituals of the Hamatsa—their insights are in fact ambiguous. However, the confrontation with ambiguity and vagueness can also be regarded as a method. The game players are consciously immersed in situations that are designed not to reveal themselves to them fully. For example, a villager explains to Jacobsen that he used to be called Raven but that he is now called Miller

Fig. 7.7 Screenshot of *Totem's Sound*, Chap. 1 (Gold Extra 2014, http://www.totems-sound.com/, last accessed January 6, 2017)

(Fig. 7.7). Later in the story, we are told to "become the Raven" or that "the Raven is still flying." The background of the dialogues surrounding "the Raven" is not clarified throughout the entire game.

Jacobsen further encounters mysterious moments. He undergoes various tests that are not explained and that he (and the game players) more or less accept by simply playing the game. At one point, Jacobsen and the game players are literally left in the dark while walking through a forest. The computer screen turns black to resemble the feeling of being lost. These moments are not simply results of uncertain factual information in Jacobsen's journey, but they are a method to reveal that ethnographers cannot know and explain everything and that their field-work entails the acceptance of uncertainties that may later translate into incomplete or lacking documentations of their encounters. The game's narrative thereby plays with perceptions and premises of knowledge and employs ambiguity and vagueness as tools to bridge information gaps in introductions of unknown time.

Ambiguity can have a positive and a negative effect. It points to the real lack of information and to the question whether museums can and should explain everything. However, by leaving out the 'wrong' context, ambiguity can also spark a renewed fascination with the exotic, encourage cultural categorisation, or legitimise generalisation. As already mentioned, the cannibalism motif that is evoked in the sequence about the biting ritual employs one of the most exotified cultural practices without ironically questioning the European obsession with it. The representation of uncertainty and ambiguity as facts in ethnographic documentation should therefore be linked to an employment of uncertainty as a method to point visitors to Europe's obsessive longing for artefacts and logical explanations in the nineteenth century.

Conclusion: Narrative Strategies of Introducing Unknown Time

Common narrative strategies used in both case studies to introduce unknown time include the use of multi-perspectivity or subjectivity, the employment of irony, as well as the demonstration of ambiguity and uncertainty with regard to ethnographic writing and documentation. These narrative strategies have two main effects that I shortly summarise in the following.

First, they create a noticeable distance between history and the representation of history. Neither case study makes a claim to completeness or truth. *Totem's Sound* emphasises its inspiration by Johan Adrian Jacobsen's travelogue but leaves no doubt that most parts of the story are fictional. By means of the ironic and often ambiguous dialogues and the internal focalisation, it is clear from the beginning that the information in the story is just one interpretation or appropriation of historical events. *Foreign Exchange* equally achieves this distance, albeit by different means. The contemporary artists featured in the exhibition reflect upon specific items of the collection from an individual perspective. These perspectives are either described in the catalogue or remain for the visitor to interpret. The juxtaposition of contemporary and ethnographic pieces instigates an artistic and therefore explorative dimension that does not aim at constructing a coherent historical narrative. Both exhibition designs hence employ rhetoric and artistic methods to avoid a descriptive and objective portrayal of history. Irony, ambiguity, and multi-perspectivity then serve as means to overcome and contrast traditional narratives of otherness that present Europe as "the brain of the Earth's body" (Preziosi 1996). They are also effective in countering the evocation of a positive fascination in the form of an imagined nostalgia for the unknown time of collection history: this is particularly important as much of the material in the archives of ethnographic museums is meant to spark a fascination with an exotic Other. As Benedikte Bjerre's work *Google Image Search of 'Bow Holder'* shows, an ironic appropriation of such visual rhetoric exposes the ideology behind it and immediately counters it by equally visual means. Finally, measures of irony, multi-perspectivity, and ambiguity can become means of making transparent the uncertainties that museums are facing with regard to introducing unknown time. The disclosure of institutional debates about collection history can help reestablish a public awareness of museum curators and exhibition designers as individual and active agents.[1]

Second, narrative strategies of multi-perspectivity/subjectivity, irony, and ambiguity serve as distancing devices between the museum's current workings and the disputed practices of collectors in the nineteenth century. Although museums are increasingly willing to make transparent their collectors' involvement in colonial exploitation, these stories can only be told so long as they are firmly located in the past. Both case studies depict colonial history as a bygone time. The works in *Foreign Exchange* are openly critical of the practices of the museum's founding director Bernhard Hagen. However, through the juxtaposition of material from the collection with contemporary art, the exhibits emphasise their own as well as the museum's detachment from the past. In general, the exhibition represents a space for discussion of nineteenth-century practices rather than a re-thinking of current institutional practices, responsibilities, and prerogatives of interpretation. Similarly, the criticism in *Totem's Sound* remains on a general, relatively decontextualised level. Johan Adrian Jacobsen is portrayed as an individual. His activities, as ironic and ambiguous as they may be, are not connected to current museum and ethnological practices. A discourse on the legitimacy of the

---

[1] Fraser and Coulson also suggest this in the context of label writing in the museum, where the disclosure of the curator's agency would equal "a direct connection between a storyteller and their audience that stimulate[s] active, imaginative engagement" (2012: 225).

ethnographic museum as institution is hence neither built into nor meant to be sparked by the narrative.

In both examples, the value systems underlying ethnographic museums today thus remain largely unquestioned, which denies the visitor an understanding of the effects that ethnographic practices and ideas in the past have had on those of the present. What is achieved by means of introducing unknown time, however, is a questioning and reflection of knowledge processes in both research and museum practice. Thereby, the museum's ambiguity and uncertainty as opposed to its omniscience becomes acceptable to some extent. Irony, ambiguity, as well as subjectivity and multi-perspectivity, enable museums not only to address a factually or strategically unknown time but also to overcome their former role as solemn narrators of truth. Museums can use these narrative strategies to open up discursive frameworks in which visitors will not necessarily acquire more knowledge about a certain time but may challenge what they already accept as true. As a result, unknown time could be regarded not as a problem but as a method: if any time can be represented as unknown by displaying ambiguities and juxtaposing different historical perspectives, why not use ambiguity as a starting point for reimagining the museum's abilities and potentialities to shape the meaning of knowledge?

BIBLIOGRAPHY

Appadurai, Arjun. 1996. *Modernity at Large: Cultural Dimensions of Globalization.* The Public Worlds Series, vol. 1. Minneapolis: University of Minnesota Press.

Bal, Mieke. 1992. Telling, Showing, Showing off. *Critical Inquiry* 18 (3): 556–594.

Burn, Andrew. 2014. Role-Playing. In *The Routledge Companion to Video Game Studies*, ed. Mark J.P. Wolf and Bernard Perron, 241–250. New York: Routledge.

Buschmann, Heike. 2010. Geschichten im Raum: Erzähltheorie als Museumsanalyse. In *Museumsanalyse: Methoden und Konturen eines neuen Forschungsfeldes*, ed. Joachim Baur, 149–170. Bielefeld: Transcript.

Centenary Gallery. *The Horniman Public Museum and Public Park Trust.* http://www.horniman.ac.uk/visit/displays/centenary-gallery. Accessed 13 Jan 2016.

Deliss, Clémentine, and Yvette Mutumba (eds.). 2014. *Ware & Wissen, or the Stories You Wouldn't Tell a Stranger.* Zürich: Diaphanes.

Clifford, James. 1988. On Collecting Art and Culture. In *The Predicament of Culture: Twentieth-Century Ethnography, Literature, and Art*, 215–252. Cambridge: Harvard University Press.

Die Welt in der Vitrine: Museum. *Rautenstrauch-Joest-Museum Online*. http://www.museenkoeln.de/rautenstrauch-joest-museum/default.aspx?s=109. Accessed 13 Jan 2016.

Fabian, Johannes. 1983. *Time and the Other: How Anthropology Makes Its Object*. New York: Columbia University Press.

Foucault, Michel. 1986. Of Other Spaces, trans. Jay Miskowiec. *Diacritics* 16: 22–27.

Fraser, Annabel, and Hannah Coulson. 2012. Incomplete Stories. In *Museum Making: Narratives, Architectures, Exhibitions*, ed. Suzanne MacLeod, Laura Hourston Hanks, and Jonathan Hale, 223–231. London: Routledge.

Geertz, Clifford. 1973. Thick Description: Toward an Interpretive Theory of Culture. In *The Interpretation of Cultures: Selected Essays*, 3–30. New York: Basic Books.

Gerstenblith, Patty. 2011. Museum Practice: Legal Issues. In *A Companion to Museum Studies*, ed. Sharon Macdonald, 442–456. Chichester: Wiley-Blackwell.

Giesen, Bernhard, Werner Binder, Marco Gerster, and Kim-Claude Meyer (eds.). 2014. *Ungefähres: Gewalt, Mythos, Moral*. Weilerswist: Velbrück Wissenschaft.

Gold Extra and Causa Creations. 2014. *Totem's Sound*. In the framework of Humboldt Lab Dahlem, *Probebühne 4*. Kulturstiftung des Bundes, Stiftung Preußischer Kulturbesitz, Berlin. http://www.totems-sound.com/. Accessed 6 Jan 2016.

Hein, George E. 2002. *Learning in the Museum*. Museum Meanings, vol. 2. London: Routledge.

Hooper-Greenhill, Eilean. 2000. *Museums and the Interpretation of Visual Culture*. Museum Meanings, vol. 4. London: Routledge.

Kaur Kahlon, Raminder, and Parul Dave-Mukherji. 2014. Introduction. In *Arts and Aesthetics in a Globalizing World*, eds. Raminder Kaur and Parul Dave-Mukherji, 1–21. London: Bloomsbury.

König, Viola, Andrea Rostásy, and Monika Zessnik. 2014. Presseinformation Probebühne 4, Projekt 'Reisebericht.' *Humboldt Lab Dahlem*, Kulturstiftung des Bundes, Stiftung Preußischer Kulturbesitz. www.das-helmi.de/images/HLD_PM_PB4_05_Reisebericht.docx. Accessed 6 Jan 2016.

König, Viola, Andrea Rostásy, and Monika Zessnik. Reisebericht/Projektbeschreibung: Neue Erzählformate für Sammlungsgeschichten. *Humboldt Forum Online*, Humboldt Lab Dahlem. http://www.humboldt-forum.de/humboldt-lab-dahlem/projektarchiv/probebuehne-4/reisebericht/projektbeschreibung/. Accessed 13 Jan 2016.

Kraus, Michael. 2015. Abwehr und Verlangen? Anmerkungen zur Exotisierung ethnologischer Museen. In *Quo Vadis, Völkerkundemuseum. Aktuelle Debatten zu ethnologischen Sammlungen in Museen und Universitäten*, ed. Michael Kraus and Karoline Noack, 227–256. Bielefeld: Transcript.

Preziosi, Donald. 1996. Brain of the Earth's Body: Museums and the Framing of Modernity. In *The Rhetoric of the Frame: Essays on the Boundaries of the Artwork*, ed. Paul Duro, 96–110. Cambridge Studies in New Art History and Criticism. Cambridge: Cambridge University Press.

Sturge, Kate. 2006. The Other on Display: Translation in the Ethnographic Museum. In *Translating Others*, ed. Theo Hermans, 431–439. Manchester: St. Jerome.

CHAPTER 8

"God's Time Is the Best:" The Fascination with Unknown Time in Urban Transport in Lagos

Daniel E. Agbiboa

INTRODUCTION

Everyday life can be regarded at once as the most self-evident and the most puzzling of all facticity. Some regard it as that time when nothing happens. Others see it as essentially unknown because the time of the everyday escapes us; it eludes our grasp. Still others identify it with the repetitive, the negative, the residual, the taken-for-granted continuum of mundane activities (Felski 2000: 78). If by *everyday* we mean that which happens day after day, it follows that it is a temporal (time-based) term whose essence subsists in the endless cycle of repetition. For this reason, it has become an *idée reçue* to deprive everyday life of critical reflection and the capacity for transcendence (Schutz and Luckmann 1983: 106–148). Notable spatial theorists such as Henri Lefebvre go further to identify the quintessential structure of everyday life with the nonaccumulative and remorseless routine of "cyclic time" (2002: 48) (immanence, or what I call 'known time'). For Lefebvre, immanent time constitutes

D.E. Agbiboa (✉)
School for Conflict Analysis and Resolution, George Mason University, Arlington, VA, USA

S. Baumbach et al. (eds.), *The Fascination with Unknown Time*, DOI 10.1007/978-3-319-66438-5_8

a drag on the permanently progressive and accumulative time of modernity, which he terms "linear time" (ibid.) (transcendence, or what I call 'unknown time'). "In the study of everyday," Lefebvre writes, "we discover the great problem of repetition, one of the most difficult problems facing us" (1987: 10).

Turning the table on Lefebvre's argument that the structure of everyday life is closely associated with the nonaccumulative routing of cyclical or immanent time whereas it lags behind the forward-moving linear or transcendent time, I argue that cyclical and linear time are in fact intertwined in lived reality and popular imagination. This concept suggests that the ebb and flow of time cannot be grasped in rigidly binary terms such as the opposition of cyclical and linear time. Interrogating popular arts such as the rhythmic use of entextualised slogans that are prominently etched on the bodies of trademark commercial minibus-taxis (*danfos*) in Lagos, Nigeria's commercial epicentre and Africa's largest city, I argue that the interaction of these seemingly conflicting representations of time affects and ultimately shapes the foundations of our meaning(lessness), (in)security, and sense of being-in-the-city. At these interfaces and interstices of conflicting notions of time, and in the interchange between (un)familiar termini, a powerful sense of unknown (or future) time can emerge, which in turn reinforces the need for a more experimental repositioning and reorientation in everyday urban life.

In the following, I first discuss the context of study and present the eclectic approach that was deployed for data collection, before focusing analytical attention on vehicle slogans as a representation of the popular fascination with, and the possibilities incumbent within, unknown time. The chapter shows how imaginations or conjectures about the future are critical for understanding the here and now.

Riding on Slogans: Lagos and the Interface Between Time

Transport is a vital resource for practising urban space and, thus, appropriating it. As a "basic principle of modernity" (Canzler et al. 2008: 3), and, hence, of the linear, transport is often identified with "liberty" and "progress" (Cresswell 2010: 21). For a large number of Africans, however, transport is essentially an "emotional, relational and social phenomenon captured in the complexities, contradictions and messiness of their everyday realities" (Nyamnjoh 2013: 653). In Lagos, for example, transport is a significant variable in asymmetrical power relations (Agbiboa 2016), stoking inequalities, marginalisation, and disconnections in everyday life. My argument therefore integrates a "rhythmanalysis" (Lefebvre 2004) of transport because of its immanent and transcendent timescapes[1] and its social embeddedness in the rhythms and vortexes of cities. As Lefebvre argues, "everywhere where there is interaction between a place, a time and an expenditure of energy, there is rhythm" (ibid., 15). Following Lefebvre, therefore, rhythmanalysis is primarily concerned with the rhythms that manifest themselves in our surrounding sense-world and what they tell us about it. Every rhythm, of course, implies a "relation of a time to a space, a localised time, or if one prefers, a temporalised space" (ibid., 89). As such, time is woven into the fabric of public space and, thus, into the everyday rhythms of the occupiers of space. Invariably, public space, imagined here as a multiplicity of publics and counter publics, becomes 'spaces of representation' (Lefebvre 1991) where encroachments, intrigues, contestations, and negotiations abound with a visible regularity. Such fluidity illuminates "the dynamics of domesticity and familiarity, [which] inscrib[e] the dominant and the dominated within the same *episteme*" (Mbembe 2001: 110 [emphasis in original]).

If the object of rhythmanalysis is transport and mobility, my concrete empirical background is Lagos, a collection of islands that are separated by creeks and the lagoon, with bridges connecting the islands to the mainland. Given the daily "transductive flows" of this place (Weate and Yusuf 2003: 19), Lagos represents rhythm par excellence. The precarious micropolitics of this uniquely dense place, combined with the abiding awareness that "time is money," compels its poorest and most desperate denizens to dwell "24 hours on the road" because there is "no time to check time" (see Fig. 8.1). These cited slogans, displayed on the bodies of *danfos* in motion, suggest that time, being a productive resource—in other words, a "carrier of significance, a form through which we define the content of relations" (Fabian 1983: ix)—necessitates prudent management for maximum achievement of livelihood and welfare ethics. Time expenditure thus furnishes a framework for making sense of the

---

[1] The word *timescape* derives from the word landscape, but instead of placing the emphasis on the visible and the spatial, as the word landscape implicitly can be said to do, timescape refers to the rhythms or temporalities that prevail in a given area or situation (Svenstrup 2013).

Fig. 8.1 "No Time to Check Time." All pictures in this chapter were taken by the author during fieldwork in Lagos, Nigeria

dynamics of poverty management among the urban poor (Yunusa 2005: 195), especially their tenacious drive to maximise profit and minimise risks in a "24-hour society" (Moore-Ede 1993; Kreitzman 1999) where people are usually on their "last warning", as one slogan puts it. In this growing megacity of Lagos that knows no inertia, the interminable struggle to impose productivity and predictability on time produces an overloaded city:

> There is overloading: overloading of language, overloading of public urban transport, overloading of living accommodations. ... Everything leads to excess, here. There is the noise of car horns, the noise of traders seeking to 'fix' a price, the noise of taxi drivers arguing over a passenger, the noise of a crowd surrounding quarrelling neighbors. There is the infernal noise of music from discotheques and bars. All this overloading constitutes an aspect, not of the environment, but of the culture itself. (Mbembé 2001: 147)

No wonder urban planners tend to describe Lagos as "the interface between time and the interchange between destinations; the meeting space for people between places, the living stage where a collage of scenes are acted and played out without a script" (Aradeon 1997: 51).

Perhaps, nowhere better typifies the rhythm or interface between time in Lagos than motor parks, bus stops, junctions, and highways where fieldwork for this chapter was conducted (specifically in Oshodi and Alimosho local government areas). These locations are not only overloaded spaces where the powerful energies of daily life in Lagos—creative, malevolent, and ambiguous—converge, but also where governance is practiced and contested. When I was not "hanging out" (Woodward 2008) in and around these transit spaces, I joined Lagosians by the roadside, which was always overloaded by the appropriation of every inch by a vibrant informal commerce amidst poor urban planning. With no start-up loan to rent a retail space, mobile street vendors set up collapsible stalls by the roadside and itinerant hawkers often cash in on the extreme traffic congestion ("go-slows," as they say in Lagos) to adroitly weave in and out of traffic, hawking all of daily life's necessities. In short, "walking in the city" (de Certeau 1984: 91) of Lagos is very much an 'art' of improvisation, which constitutes "a determining element of behavior and urban knowledge" (Mbembé 2001: 147).

Foregrounding overload as context in Lagos, my approach focuses on the city's commercial *danfos* while using their rhythmic and colourful art slogans as an analytic lens that helps identify the ways in which Lagosians carve out meaningful temporalities and define their aspirations for the 'city yet to come' (Simone 2004). The ubiquitous nature of these slogans suggests that they constitute the occupational sub-culture of urban transport, imbuing everyday life and social relationships with meaning and purpose (Agbiboa 2016). During my fieldwork, I collected 312 vehicle slogans from the rickety bodies of *danfos* (moving and stationary) on Lagos roads. The content analysis of these slogans was informed by interviews with operators, owners,[2] and regular commuters in Lagos. These interviews were then supplemented by observations drawn from my long residence in Lagos and my experience as a *danfo* user, gazing upon these vehicle slogans on a daily basis. Some of the slogans were so cryptic in meaning that it was only by directly questioning individual operators about their unique choice of slogans that I could discern "the text within the text" (Lotman 1988), thereby bypassing the "misleading one-sidedness of textual interpretation" (Jaworski and Thurlow 2010: 15). However, in some instances, I was unable to do this because I collected the slogans from *danfos* in perilous motion.[3] Although this may go down as a study limitation, it is noteworthy that lexico-semantic meanings are fluid and not entirely under the author's control. As Barber argues, "texts generate 'surplus': meanings that go beyond, and may subvert, the purported intentions of the work" (1987: 4). So understood, I derived intersubjective meaning from the interpretations offered by other operators in Lagos (Fig. 8.2).

Slogans are imagined here not only as abstract and discursive but "embodied, felt, interactive and cumulative" (Morgan 2008: 228) archives of urban life and spatial performance. Hence, slogans mirror the "simultaneous promise, threat, and resource" of the city (Simone 2010: 3). In terms of their form and content, most slogans are typically short phrases that derive their meaning(s) simultaneously from repertoires of orality and literacy (Quayson 2014: 130). Examples of slogans include

---

[2] It is not unusual for *danfos* to change ownership several times during their lifespan with subsequent owners keen to impose their own personal slogans on newly procured *danfos*.

[3] In Lagos, as in many African cities, a combination of high speeding and driving 'on the edge' is required if workers are to recoup the daily payment to the owner as quickly as possible. The driver must race between two end points of his chosen route at the highest speed he can possibly generate, while his conductor calls out destinations at bus stops or anywhere he spots passengers.

Fig. 8.2 "Last Warning."

"Trust," "Let Them Say," "No Shaking," "Ghetto Boy," and "*Aiye Mojuba*" (I respect life). A number of slogans draw on local proverbs ("*Ise l'oogun ise*" [Work is the cure for poverty]), holy texts ("No Food for Lazy/Idle Man"), or street slang ("*Jeun Soke*" [Eat Upwards]).[4] A great many slogans are multilingual, using a creative blend of the linguistic art form of Nigerian Pidgin English or the dominant lingua franca in Nigeria (Yoruba, Hausa, and Igbo). For instance, the slogan "*wa-zo-bia*" appropriates the Yoruba, Hausa, and Igbo[5] words for *to come*: this is the *danfo* owner's way of saying 'all are welcome' in his bus. Thus, in some sense, slogans are "operations of the productive imaginations" (Mbembé 2001: 159). The inherent diversity of slogans reaffirms the plurality of Lagos life, a 'no man's land' where myriad cultures and variegated subjectivities intersect and produce cultural ripples across Nigeria and beyond. Finally, slogans share an ethos with local proverbs, not only because of their short and witty nature, but because of their "entextualization"—the process by which circulable texts are (re)produced by extracting discourse from its in situ context (Bauman and Briggs 1990: 72–78).

THE POSSIBILITY OF UNKNOWN TIME

"God Moves with His Own"

The transport industry in Lagos is marked by insecurity on many levels. Many *danfos* ply without licenses and flout traffic rules with impunity. These *danfos* are generally perilously overloaded and dangerously speeding. As such, *danfos* have an unenviable reputation for causing extreme go-slows and ghastly accidents on the road. The situation is complicated by the fact that many operators drive under the strong influence of *paraga*, a local beverage with high alcoholic content. In 2013, the Lagos State Drivers' Institute (LASDRI) conducted a test on *danfo* drivers and found that 22% of them are "partially blind" (Akoni 2013); the results of another survey in 2015 showed that more than 99% of *danfo* drivers are "hypertensive" (Akinola 2015). Yet, many inhabitants of the city continue to use *danfos* each day *faute de mieux*. As one commuter in Oshodi said to me: "We know most of the *danfos* are death traps but since

---

[4] For more vehicle slogans, see appendix.

[5] The three majority ethnic groups in Nigeria.

we can't afford the high taxi fares, we have no choice but use them."
During the course of my fieldwork, I sighted many rickety *danfos* (with
worn-out tyres and no functioning indicators) displaying slogans such
as "Trust in God Always," "Let Us Pray," "Relax! God is in Control,"
"God Moves with his Own," "Remember Now Thy Creator," and "Pray
and Hope." In appealing to a divine force through slogans, these operators have found a way of reassuring worried commuters that they are in
safe (divine) hands, despite the bad conditions of the *danfo*. Here, we see
how a transcendental being (God) is used to allay the existential fears of
commuters and to help them transcend the public knowledge that *danfos*
reminds you of 'Ur Six Feet' (death).

"No Loss. No Lack. No Limitation"

The struggle for survival and security writ large remains the overwhelming concern of many operators in Lagos. As one driver recounted: "To
eat, you need to hustle everyday because stomach has no holiday." The
'hard life' of these ragtag operators is not helped by the presence of
motor-park touts on the road, a dreaded cohort of 'youthmen' locally
known as *agberos*. These marauding *agberos* represent the interests and
extortionate power of the National Union of Road Transport Workers
(NURTW), the most politicised and violent trade union in Nigeria
(Albert 2007), and are employed to collect illegal taxes from operators
daily. Usually acting under the influence of *paraga*, *agberos* tend to vio-
lently attack operators who delay in parting with their money. During
fieldwork, for example, I witnessed the violent death of four operators in
the hands of *agberos*—nicknamed the 'kings of Lagos roads.' Reflecting
the negative sociopsychological dispositions of operators are slogans such
as "*aiye le*" (The world is hard), "*aiye nika*" (Wicked world), "*aiye daru*"
(The world has no order, no morality sense), "*aiye toto*" (Beware of the
world), and "*aiye ore eke*" (Life, friend of deceit). Here, everyday events
are perceived as speckled by devilish forces, both physical and spiritual,
working against one's struggles to "become somebody important in life"
("*eniyan Pataki*").

For many operators, "everyday life has come to be defined by the par-
adigm of threat, danger, and uncertainty. A social world has gradually
taken form where general distrust and suspicion go hand in hand with
the need for protection against increasingly invisible enemies" (Mbembe
2006: 310). Hence, operators, especially those who own *danfos*, use

their slogans as a talisman against *awon aiye* (the world),[6] as a warning to jealous *ota* (enemies), and as a prayer for "No Loss. No Lack. No Limitation."[7] In this respect, slogans such as "Back to Sender" express the owner's supplication that any bad wish towards his business reflects back to those who wish them. Others such as "Sea Never Dry" reflect the owner's wish that his *danfo* (his source of income) will never leave the road. Implicit in slogans like "No Weapon Fashioned Against Me Shall Prosper," "Touch Not My Anointed," "Blessed from Above Is Above All", "*Iwo dan wo*" (You dare), or "Blood of Jesus" is a warning of the divine power that protects the owner from unknown harm. In this case, the message to the *ota* is clear: if you take me on, you take on God himself. In line with de Certeau, these slogans may be seen as "art[s] of the weak" and a "tactic" (1984: 37) for negotiating and navigating an everyday urban life of innumerable interruptions and interference. "The tactic," argues de Certeau, "*depends on time*—it is always on the watch for opportunities that must be seized 'on the wing.'... It must constantly manipulate events in order to turn them into 'opportunities'" (ibid., xix [my emphasis]).

"No Condition Is Permanent"

Arjun Appadurai once argued that the poor's aspirations, nonhomogeneous as they appear, are "inevitably tied up with more general assumptions about the good life, and life more generally" (2004: 68). According to Appadurai, "the poor ... express horizons in choices made and choices voiced, often in terms of specific goods and outcomes" (ibid.). This attitude is true of the poor transport operators in Lagos. Despite their hard life (which many *danfo* drivers and conductors described in terms

---

[6]Whenever a *danfo* slogan contains *awon aiye*, it, more often than not, evokes various reactions, ranging from threat to acknowledgement of a malevolent force that can do great harm to one's *ori* (destiny). *Aiye* (the world) is a power, a superior that must be constantly supplicated (*aiye e ma binuwa* [World don't be angry with us]) and respected (*aiye mojuba* [I respect the world]) by owners lest their business be altered (*aiye e ma pa kadara* [World, don't alter our fate]). Many owners that I spoke to also used *aiye* to describe operators who are cunning, dishonest, unpredictable, and uncertain in their everyday social relationships.

[7]The fears of transport owners are not surprising if we bear in mind that in Lagos, and many parts of urban Africa, vehicle ownership comes with a potent mix of prestige, envy, and contempt (Agbiboa 2016).

of 'war'), I was always struck by the fact that many operators never ceased to dream about a better life, including owning a *danfo*, good health, travelling to *obodo oyibo* (land of the white man), winning the American visa lottery, marrying a graduate, becoming *eni olokiki* (someone famous), *olowo* (someone rich), *eni giga* (someone in high places), *eni pataki* (someone important), paying for one's children's education, acquiring a diploma, and so on. These varied dreams "are not simple fantasies woven from sleep [but] … a normal technique for solving a problem or *finding a way* out of a dilemma" (Burridge 1995: 219 [my emphasis]). Thus, the dream is the chosen vehicle for expressing, if not reclaiming, the hopes of poor transport workers in a precarious business environment where 'anything can happen.' As Burridge remarks:

> When expressed through a dream hope merges into positive expectancy. Any man may wish or hope for something at any time. But, when in association with a dream the hope comes near to realization. Dreams tend to pull a future into current sensible reality; they give definity to hope, adding faith, thereby putting the dreamer in touch with a verity shortly to be manifest. (ibid., 180)

In the face of an urban life that seems to be falling apart before their own eyes, few operators that I encountered accept "victim identities" (Kihato 2007: 413) which analytically would alienate them from their social agency in ways that perpetuate imageries of disempowerment. Rather, these poor operators use positive qualifiers—that is, temporally inflected and hope-imbued slogans—to manage the radical uncertainty and bankruptcy of the present, the here and now. Such future-looking slogans include these:

> "One Day Is One Day," "I Won't Stay This Way for Ever," "God's Time Is the Best," "All in God's Time," "God *Dey*" (There is God), "The Downfall of a Man Is Not the End of His Life," "My Beginning Small, My End Great," "A Patient Dog Eats the Fattest Bone," "No Competition in Destiny," "The Storm Is Over," "No Condition Is Permanent," "Time Will Tell," "The Young Shall Grow," "You Will See," "It Will Be Well," "Very Soon," "One Day," "*Ko ni baje*" (It will not spoil), "No One Knows Tomorrow," "I Shall Return."

The positive expectancy incumbent within these slogans is epitomised by the official Lagos slogan, "*Eko o ni baje*" (Lagos will not spoil). Or as one *danfo* driver modified it, "*Eko o ni baje ju bai lo*" (Lagos will not spoil more than this). Slogans—such as "One Day"—reinforce not only the temporality of hope (Crapanzano 2003) but also the fact that "conjectures about the future [form] an implicit part of the understanding of the present" (Moore 1987: 727). In this regard, the popular slogan "God's Time Is the Best" is one of the ways in which operators navigate a highly competitive and uncertain business where "fortunes are seen to be made and lost and one's own fortune often appears to be beyond one's control" (Williams 1980: 114). Thus, unknown time is meant to liberate the operator from the known and familiar (dis)order of everyday urban life. Put differently, unknown time serves as a point of departure or escape from a current terrain of known experiences that poses the greatest danger or risk to the self-fulfilment and aspirations of operators.

Ultimately, therefore, the operator's capacity to aspire, to hope for a better tomorrow, reveals their abiding awareness that "No Condition Is Permanent" (see Fig. 8.3); only the permanence of change is unconditional. In other words, the temporal uncertainties of everyday life catapult operators into some mystical divine realm, which enables them to transform their precarious materiality into "spaces of hope" (Harvey 2000) and "active waiting" (Cooper and Pratten 2015: 11). This appeal to, and fascination with, the unknown as a vital resource for managing uncertainty is illustrated by a hymn on the walls of a motor-park in Lagos. The hymn stresses the fluid nature of day-to-day life and the power incumbent within the unknown:

> Remember that the wealthy person becomes poor. The poor person often becomes wealthy in this world. The king of a town becomes poor in this world. And the poor may become a mighty king. (Field notes, November 16, 2014)

Granted that the hopes of operators are premised on the very fact of uncertainty ("No Condition Is Permanent"), it follows that "to seize the opportunities that luck or fate may bestow requires anticipation and mobility—a propensity to being in the right place at the right time" (Cooper and Pratten 2015: 11). In this respect, being young and surviving on the margins of the economy is recast as still having time and still having hope. This meaning is conveyed in popular *danfo* slogans such as "Young Shall Grow," "It's Great to Be Young," "Today Work, Tomorrow Success," and "Better Is Coming." In the light of these

Fig. 8.3 "No Condition is Permanent."

temporal horizons,[8] uncertainty produces dispositions by which opera-tors not only get by in the present but also get on in the future. This realisation led Johnson-Hanks to intend "a unit of social analysis based in hope rather than event" (2002: 865–866). Jane Guyer has already called for an ethnography of the "near future" which arises from a concern for empirically informed and more open-ended studies of processes of change that interrogate "the reach of thought and imagination, of plan-ning and hoping, of tracing out mutual influences, of engaging in strug-gles for specific goals" (2007: 409). What arises from an analytical focus on the "near future" is a better appreciation of how the urban poor tacti-cally position themselves and stay vigilant to future possibilities, against all odds. This attitude relates to what Appadurai calls the "ethics of pos-sibility" (2013: 299)—that is, ways of thinking, feeling, and acting that increase hopeful horizons.

The Turn to "Juju"

Against the background of an everyday urban life where 'anything can happen,' many poor operators expressed conviction in some form of divine intervention needed to enhance their success and safety on fam-ished roads. Many that I spoke to turned to *juju* (charms) as a tactic against accidents and bad luck (i.e., unknown enemies working against one's good) as well as to attract passengers and to ensure self-protec-tion and advantage in an extremely competitive transport business. The appropriation of *juju* by operators echoes Michael Taussig's illuminating work on *The Devil and Commodity Fetishism*[9] *in South America* (1980), which depicted sorcery as resistance to the numerous material and ideo-logical contradictions fostered by capitalism, a creative assertion of the human will. Taussig found that poor Bolivian miners and Columbian plantation workers regularly invoked the devil to secure production of tin or sugarcane (1980: 3). He studied the "preexisting cosmogony" of the peasants as a source of vital images, beliefs, and rituals, which the people reinterpret to meet current needs (ibid.). In both instances—the

---

[8] Horizons are time specific: "what looks like a hopeful prospect now may be shut down without warning tomorrow, and another potential future may open up" (Johnson-Hanks 2002: 872).

[9] This term is used here to refer to "the adventurous appropriation of an exotic object as a means of increasing one's own status and value" (Weate and Bakare Yusuf 2003).

miners in South America and the operators in Lagos—we see how hard-pressed informal actors tendentially appeal to the power of the transcendental or the unknown to manage recurrent events of turbulence impose predictability on their everyday life. Such practices do not only reduce the anxieties of these workers or give them (the illusion of) control over a precarious situation in which they are immobilised, they are also ways to make sense of the mysterious world in which these informal workers weave their existence. The *juju*s used by operators are, however, not substitutes for laziness. On the contrary, they are deployed to "facilitate profit and protect the individual and property in a competitive environment" (Lawuyi 1988: 6).

Iris Young has already argued that "dwelling in the world means we are located among objects, artefacts, rituals, and practices that configure who we are in our particularity" (1997: 153). This feeling is true of the many *juju* that caught my eyes during my frequent trips using *danfos* in Lagos. The *juju* comes in various forms, including the shape of a lock or a comb, a dry head of a rat tied to a cowrie, a dry animal skin twisted into a rope and tied around the driver's arm (attached to it are cowries), or a feather with a red rope tied to it. These symbols—often hidden under the dashboard, tied to the mirror, or placed under the driver's seat—display "a form of empowerment that expresses 'the fact of powerlessness'" (Gelder and Thornton 1997: 375). A study of Yoruba taxi drivers in the 1980s, for example, found that 80% of Muslims and 60% of Christians had protective charms in their vehicles, a "resort to symbolic action in the face of uncertainties" (Lawuyi 1988: 4). In Lagos, Lawuyi argues:

> Armed robbers may attack and steal the vehicles [i.e., the symbol of livelihood]. The roadside mechanic may have mistakenly connected two wires that could ignite and burn the vehicle. Nobody can predict when accidents will occur, as drivers in a hurry overtake dangerously. *The juju in the vehicle serves to remind the drivers of their power to escape from any danger.* ... To these symbols ... are ascribed the power to make drivers disappear into thin air when accidents occur. Alternatively they may prevent accidents. They also bring wealth by attracting passengers. Through their mystification the juju make it possible for the individual to survive the disruptive processes created by national and international tensions, which affect economic processes. ... The juju themselves are not mechanical reflections of power relations, but are autonomous entities which can act on the power order and modify it. (ibid. [my emphasis])

Notably, I spoke to a group of drivers who related their habitual offering of weekly food sacrifices to *eshuona* (the dreaded god of the road) before embarking on their business. According to these operators, their sacrifices are intended to propitiate *eshuona* so as to fortify themselves against his tricks and to implore him to *fi wa le* (leave us alone). Here, repetition is not necessarily tantamount to the operators' domination or enslavement in the ordinary, his association with immanence rather than transcendence, as implied in the work of Lefebvre (1987). Quite on the contrary, these repetitive (weekly) sacrifices help operators to transcend their precarious materiality and avert any danger in the nick of time. In this case, repetition points to "resistance through rituals" (Hall and Jefferson 1993) and "ritualistic and symbolic opposition to a dominant order" (Highmore 2002: 12). This statement reinforces Rita Felski's argument that "repetition is one of the ways in which we organise the world, make sense of our environment and stave off the threat of chaos" (2000: 21).

CONCLUSION

This chapter has argued that linear (transcendent) time is not at odds with the cyclical (immanent) time of everyday life. In contrast to some commentators who approach everyday life as merely repetitive and bereft of transcendence or critical reflection, I found the capacity for transcendence to be central to the routine lives of transport operators in Lagos. I also found that the uncertainty of linear time imbues operators with the capacity to aspire, the audacity to hope, and the tenacity to survive against all odds. I was repeatedly struck by the way in which, in the popular imagination of transport operators, the future coexisted with the present (but was not coextensive with it); the supernatural manifested in the physical; and the transcendent permeated the mundane. This idea, perhaps, not only underlines the spatiotemporal embeddedness of daily life, but calls for a rethink of Lefebvre's (2002) view of the everyday as a terrain of conflict and struggle between enduring forms of cyclical time and the linear. Finally, this chapter marshals evidence from the popular arts slogans to illustrate how the urban poor manage uncertainty and assert their agency in a precarious and unpredictable business. In other words, a study of unknown time in urban transport suggests that the present and the future are not disconnected horizons of social practice but combine to make place and produce life in the city. What emerges from these blurred boundaries is a reclaiming of *the real* as "process" and *the possible* as "everything that is only partially conditioned, which has not yet been fully or conclusively determined" (Bloch 1954: 17). Hence, for poor operators who struggle each day to make the most of their time and define their hopes for a better future, the objective, as Simone argues, is

> not to tie themselves down to prevailing notions about what can be taken into account, what makes sense, or what is logically possible. The idea is to keep things open, keep things from becoming too settled or fixed. The messed-up city, then, is not simply a mess. In the very lack of things seeming settled, people keep open the possibility that something more palatable to their sense of themselves might actually be possible. (2010: 260–261)

The foregoing discussions show how the fascination with unknown time is a powerful way in which social actors in Lagos negotiate the present uncertainties and insecurities of daily life and enact future possibilities. Perhaps this calls for more scholarly inquiries into the nexus between contingency and futurity in urban life and popular imagination.

APPENDIX: LIST OF VEHICLE SLOGANS

1. 24 Hours on the Road
2. A Patient Dog Eats the Fattest Bone
3. *Aiye le* (The world is hard)
4. *Aiye Mojuba* (I respect life)
5. All in God's Time
6. Anything Can Happen
7. *Awon aiye* (The world)
8. Back to Sender
9. Better Is Coming
10. Blessed from Above Is Above All
11. Blood of Jesus
12. *Eko o ni baje* (Lagos will not spoil)
13. Ghetto Boy
14. God *Dey* (God exists)
15. God Moves with His People
16. God's Time Is the Best
17. I Shall Return

18. I Won't Stay This Way for Ever
19. *Ise l'oogun Ise* (Work is the cure for poverty)
20. It Will Be Well
21. It's Great to Be Young
22. *Iwo dan wo* (You dare)
23. *Jeun Soke* (Eat up)
24. *Ko ni baje* (It will not spoil)
25. Last Warning
26. Let Them Say
27. Let Us Pray
28. My Beginning Small, My End Great
29. No Competition in Destiny
30. No Condition Is Permanent
31. No Food for Lazy Man
32. No Loss. No Lack. No Limitation
33. No Man's Land [Referring to Lagos]
34. No One Knows Tomorrow
35. No Shaking (Feel No Fear)
36. No Time to Check Time
37. No Weapon Fashioned Against Me Shall Prosper
38. One Day
39. One Day Is One Day
40. Pray and Hope
41. Relax! God Is in Control
42. Remember Now Thy Creator
43. Remember Ur Six Feet
44. Sea Never Dry
45. The Downfall of a Man Is Not the End of His Life
46. The Storm Is Over
47. The Young Shall Grow
48. Time Is Money [originally Time na Money]
49. Time Will Tell
50. Today's Work. Tomorrow's Success
51. Touch Not My Appointed
52. Trust
53. Trust in God Always
54. Very Soon
55. *Wa-Zo-Bia* [Yoruba, Hausa, and Igbo word for *to come*]
56. You Will See

BIBLIOGRAPHY

Agbiboa, Daniel E. 2016. 'No Condition is Permanent': Informal Transport Workers and Labour Precarity in Africa's Largest Megacity. *International Journal of Urban and Regional Research,* 40: 936–957.

Akoni, Olasunknami. 2013. 14,300 Lagos Drivers Are Partially Blind—LASDRI Boss. *Vanguard.* September 17. http://www.vanguardngr.com/2013/09/14300-lagos-drivers-are-partially-blind-lasdri-boss/. Accessed 2 June 2016.

Akinola, Femi. 2015. 99% of Danfo Drivers in Lagos Hypertensive—Commissioner. *Daily Trust.* December 11. http://www.dailytrust.com.ng/news/lagos/99-of-danfo-drivers-in-lagos-hypertensive–commissioner/123799.html. Accessed 7 June 2016.

Albert, Isaac Olawale. 2007. Between the State and Transporter Unions: NURTW and the Politics of Managing Motor Parks in Ibadan and Lagos, Nigeria. In *Gouverner les villes d'Afrique: État, gouvernement local et acteurs privés,* ed. Laurent Fourchard, 125–138. Paris: Karthala.

Appadurai, Arjun. 2004. The Capacity to Aspire: Culture and the Terms of Recognition. In *Culture and Public Action,* ed. Vijayendra Rao and Michael Walton, 59–84. Stanford: Stanford University Press.

———. 2013. *The Future as Cultural Fact: Essays on the Global Condition.* London: Verso.

Aradeon, David. 1997. Oshodi: Replanners' Options for a Subcity. *Glendora Review: African Quarterly on the Arts* 2 (1): 51–58.

Barber, Karin. 1987. Popular Arts in Africa. *African Studies Review* 30 (3): 1–78.

Bauman, Richard, and Charles L. Briggs. 1990. Poetics and Performance as Critical Perspectives on Language and Social Life. *Annual Review of Anthropology* 19: 59–88.

Bloch, Ernst. 1954. *The Pinciple of Hope,* vol. 1, trans. Neville Plaice, Stephen Plaice, and Paul Knight. Cambridge: MIT Press.

Burridge, Kenelm. 1995. *Mambu: A Melanesian Millennium.* Princeton: Princeton University Press.

Canzler, Weert, Vincent Kaufmann, and Sven Kesselring. 2008. Tracing Mobilities: An Introduction. In *Tracing Mobilities: Towards a Cosmopolitan Perspective,* ed. Weert Canzler, Vincent Kaufmann, and Sven Kesselring, 1–11. Aldershot: Ashgate.

Cooper, Elizabeth, and David Pratten. 2015. Ethnographies of Uncertainty in Africa: An Introduction. In *Ethnographies of Uncertainty in Africa,* ed. Elizabeth Cooper and David Pratten, 1–16. New York: Palgrave.

Crapanzano, Vincent. 2003. Reflections on Hope as a Category of Social and Psychological Analysis. *Cultural Anthropology* 18 (1): 3–32.

Cresswell, Tim. 2010. Towards a Politics of Mobility. *Environment and Planning D: Society and Space* 28 (1): 17–31.

de Certeau, Michel. 1984. *The Practice of Everyday Life*, trans. Steven Rendall. Berkeley: University of California Press.

Fabian, Johannes. 1983. *Time and the Other: How Anthropology Makes its Object*. New York: Columbia University Press.

Felski, Rita. 2000. The Invention of Everyday Life. *Cool Moves: A Journal of Culture, Theory and Politics* 39: 15–31.

Gelder, Ken, and Sarah Thornton (eds.). 1997. *The Subcultures Reader*. London: Routledge.

Guyer, Jane I. 2007. Prophecy and the Near Future: Thoughts on Macroeconomic, Evangelical, and Punctuated Time. *American Ethnologist* 34 (3): 409–421.

Hall, Stuart, and Tony Jefferson. 1993. *Resistance through Rituals: Youth Subcultures in Post-war Britain*. London: Routledge.

Harvey, David. 2000. *Spaces of Hope*. Edinburgh: Edinburgh University Press.

Highmore, Ben. 2002. Introduction: Questioning Everyday Life. In *Everyday Life and Cultural Theory: An Introduction*, ed. Ben Highmore, 1–34. London: Routledge.

Jaworski, Adam, and Crispin Thurlow. 2010. Introducing Semiotic Landscapes. In *Semiotic Landscapes: Language, Image, Space*, ed. Adam Jaworski and Crispin Thurlow, 1–40. London: Continuum.

Johnson-Hanks, Jennifer. 2002. On the Limits of Life Stages in Ethnography: Toward a Theory of Vital Conjunctures. *American Anthropologist* 104 (3): 865–880.

Kihato, Caroline Wanjiku. 2007. Invisible Lives, Inaudible Voices? The Social Conditions of Migrant Women in Johannesburg. In *Women in South African History*, ed. Nomboniso Gasa, 379–419. Cape Town: HSRC Press.

Kreitzman, Leon. 1999. *The 24 Hour Society*. London: Profile Books.

Lawuyi, Olatunde Bayo. 1988. The World of the Yoruba Taxi Driver: An Interpretative Approach to Vehicle Slogans. *Africa: Journal of the International African Institute* 58 (1): 1–13.

Lefebvre, Henri. 1987. The Everyday and Everydayness, trans. Christine Levich. *Yale French Studies* 73: 7–11.

———. 1991. *The Production of Space*, trans. Donald Nicholson-Smith. Oxford: Blackwell.

———. 2002. *Critique of Everyday Life*, vol. II, trans. John Moore. New York: Verso.

———. 2004. *Rhythmanalysis: Space, Time and Everyday Life*, trans. Stuart Elden and Gerald Moore. London: Continuum.

Lotman, Yuri M. 1988. Text Within a Text. *Soviet Psychology* 26 (3): 32–52.

Makeham, Paul. 2005. 'Performing the City.' *Theatre Research International* 30 (2): 1–12.

Mbembe, Achille. 2001. *On the Postcolony*. Berkeley: University of California Press.

———. 2006. On Politics as a Form of Expenditure. In *Law and Disorder in the Postcolony*, ed. Jean Comaroff and John L. Comaroff, 299–330. Chicago: Chicago University Press.

Moore-Ede, Martin. 1993. *The 24 Hour Society: The Risks, Costs and Challenges of a World that Never Stops*. London: Piatkus.

Moore, Sally Falk. 1987. Explaining the Present: Theoretical Dilemmas in Processual Ethnography. *American Ethnologist* 14 (4): 727–736.

Morgan, David. 2008. The Materiality of Cultural Construction. *Material Religion: Journal of Objects, Arts and Belief* 4 (2): 228–229.

Nyamnjoh, Francis B. 2013. Fiction and Reality of Mobility in Africa. *Citizenship Studies* 17 (6–7): 653–680.

Quayson, Ato. 2014. *Oxford Street, Accra: City Life and the Itineraries of Transnationalism*. Durham: Duke University Press.

Schutz, Alfred, and Thomas Luckmann. 1983. *The Structures of the Life-World*, vol. II, trans. Richard Zaner and David J. Parent. Evanston: Northwestern University Press.

Simone, AbdouMaliq. 2004. *For the City Yet to Come: Changing African Life in Four Cities*. Durham: Duke University Press.

———. 2010. *City Life from Jakarta to Dakar: Movements at the Crossroads*. New York: Routledge.

Svenstrup, Morten. 2013. *Towards a New Time Culture*, trans. Peter Holm-Jensen. *Time-Culture.net*. https://time-culture.net/wp-content/uploads/2013/03/ Towards-a-new-time-culture-A4.pdf. Accessed 18 May 2006.

Taussig, Michael T. 1980. *The Devil and Commodity Fetishism in South America*. Chapel Hill: North Carolina University Press.

Weate, Jeremy, and Bibi Bakare Yusuf. 2003. Ojuelegba: The Sacred Profanities of a West African Crossroad. http://bakareweate.com/texts/ OJUELEGBA%20long%20version%20single%20spacing.pdf. Accessed 1 June 2016.

Williams, Gavin. 1980. *State and Society in Nigeria*. Idanre: Afrografika Publishers.

Woodward, Kath. 2008. Hanging Out and Hanging About: Insider/Outsider Research in the Sport of Boxing. *Ethnography* 9 (4): 536–560.

Young, Iris Marion. 1997. *Intersecting Voices: Dilemmas of Gender, Political Philosophy and Policy*. Princeton: Princeton University Press.

Yunusa, Mohammed-Bello. 2005. Life in a High-Density Urban Area: Anguwar Mai Gwada in Zaria. In *Urban Africa: Changing Contours of Survival in the City*, ed. AbdouMaliq Simone Abdelghani Abouhani, 177–205. Dakar: Codesria Books.

CHAPTER 9

Perpetual Wanderers—Timeless Heroes: Gypsies in European Musical Culture

Anna G. Piotrowska

In European culture—mainly in literature, and later cinema, but in music particularly (e.g., operas, operettas)—the stereotype of Gypsies as eternal wanderers has been fossilised as one of the most appealing images, impacting the imagination of an audience often acutely aware of the discrepancy between reality and artistic representations of unknown Gypsies. A very specific mechanism of the ambivalent treatment of Gypsies—one that has always constituted an integral part and parcel of the highly romanticised picture of Gypsies promoted in the arts—is their depiction as being attractive and unsettling at the same time. In this chapter, I discuss a sense of Gypsy timelessness, understood as a specific sense of timing in musical works, because music is fundamentally the art of time, organising our experience of time through the structures of sounds (Green 2011: 62). While addressing the issues of specific time treatment observed in musical compositions alluding to the imaginary Gypsy world, I concentrate, on the one hand, on stage works (operas, operettas) featuring Gypsy heroes and, on the other hand, on purely instrumental compositions called 'Gypsies' (e.g., waltzes, polkas,

A.G. Piotrowska (✉)
Institute of Musicology at Jagiellonian University,
Kraków, Poland

© The Author(s) 2017
S. Baumbach et al. (eds.), *The Fascination with Unknown Time*,
DOI 10.1007/978-3-319-66438-5_9

rhapsodies). I attempt to show how the principle of suddenness strengthens the sense of timelessness these works try to capture—as if in the hope of replicating the nonreproducible, intangible Gypsy character.

Regarding theatrical musical productions, different ways of portraying Gypsies as perpetual wanderers can be identified. Usually Gypsies are presented as people who appear on stage quite suddenly, without warning, and vanish equally unexpectedly. It can be argued that this might allude to the fact that the exact time when Gypsies had left the Indian Peninsula and arrived in Europe remains unknown. Indeed, for the average European, that is, someone unacquainted with the history of Gypsies, their appearance on the continent might appear as rather sudden in nature. Furthermore, as acclaimed wanderers constantly on the move, Gypsies seem to embody the timeless suspension between the ideal aim of their journey—the mystical, almost divine 'then and there'—and the mundane, very prosaic 'now and here.' Hence in various choruses featured in many Romantic operas, the Gypsy protagonists sing about their fate as homeless vagrants for whom time is irrelevant. A similar notion is expressed in numerous songs composed from the nineteenth century onward, including many pop songs that can be tagged as 'Gypsies,' that is, alluding to the Gypsy heroes and instancing a so-called Gypsy style of life.

In purely instrumental compositions, which refer to Gypsies in their titles, the figure of the idealised, imaginary Gypsy is expressed in very specific musical patterns and tempi: tempo rubato as well as accelerando (hastening the tempo) have a crucial role in creating the 'Gypsy' style. The latter further includes a variety of techniques that aim at manipulating the perception of time as well as creating a rather loose genre, which inevitably recalls the romanticised world of the exotic and underlines the fascinating and mysterious aura associated with Gypsies.

GYPSIES TRANSGRESSING BOUNDARIES OF SPACE AND TIME

Infinite wandering and the transgression of different times and spaces have become the hallmark of Gypsy existence. Biological (genetic) as well as linguistic evidence suggests that Gypsies stemmed from the Indian Peninsula (Kenrick 2010: xxxviii) but left their original place of settlement and arrived in Europe most probably before 1250.[1] Prone to speculations and assumptions, the exact time of their arrival in Europe has long remained obscure and uncertain; as such, it could be interpreted as unheralded, unprepared, even sudden. Because they were initially given various names, Gypsies could often go unrecognised, and it seems justifiable to speculate that their first appearances in different parts of Europe might have gone unnoticed.

Furthermore, once in Europe, Gypsies continued to travel. The majority of them only passed through eastern European territories on their further wanderings to the west: individual Gypsies are recorded to have appeared in Switzerland in 1414 and a dozen years later in France (Fraser 2007: 62, 69). Significant numbers of Gypsies followed in the tracks of these initial arrivals, successively relocating themselves in western Europe, escaping, among other things, slavery, to which they had been subjected in Romanian principalities, as well as the advancing Turks. By presenting themselves as Christians on a seven year pilgrimage, Gypsies initially enjoyed the support of the local population. During their journeys, they were protected from any dangers by the authorities, who issued guarantees of safe conduct and allowed them to move freely across a given territory.[2] With time, however, the true intentions of Gypsies were subject to speculation and the validity of these letters was brought into question.

Wandering Gypsies were perceived, first and foremost, as providers of entertainment—musicians and acrobats, magicians or fortune tellers, embellishing their performances with stories about their dangerous, adventure-filled journeys. Gypsy women displayed an especial ease at mastering local languages, using this ability for telling the future and reading people's palms. Yet, in Christian Europe the attitude towards entertainment remained ambivalent: musicians were perceived as the inheritors of quackery and the magic tradition, constituting a potential threat to Christianity, and the combination of musical, magical, and

---

[1] Entering Europe, Gypsies continued to move around: they were first encountered on the eastern fringes of the continent. Numerous accounts testify to the presence of Gypsies in fourteenth-century Constantinople, but as documents show, it is highly probable that they had been active in Byzantium by the end of the twelfth century. See Rochow and Matschke (1991: 253).

[2] Such letters of protection for Gypsies were issued by, among others, Sigismund, Holy Roman Emperor (1368–1437).

acrobatic elements prompted associations with devilish practices. Perhaps the uncertain situation of secular musicians in medieval Europe could have been an incentive for Gypsies to adopt the custom of music-making, ensuring them the possibility of leading a relatively independent lifestyle while allowing them to fill a market niche. This ambivalent position on the margins of the dominant society enabled Gypsies to engage in a kind of double game of seducing and enticing the public while openly rejecting their standards of life. At the same time, wandering Gypsy musicians worked as transmitters of information, spreading various fashions and modes in different places and environments. Travelling from one place to another, they not only disseminated news or trends but also worked as mediators between different cultures, between spaces and times. The uniqueness of Gypsies mainly resulted from their unrestricted adaptation, transformation, and infusion of various elements taken from non-Gypsy (e.g., gazhe) and Gypsy cultures, and from their unrestrained ability to freely circulate these ad lib melanges of ideas.

The Portrait of Gypsies in Musical Stage Works: Eternal Wanderers

Gypsies in Operas, Operettas, and Ballets: The Conception of Time

In the nineteenth century, at a moment when the novelty and freshness of historicism and medieval mystery as propagated by Walter Scott was wearing off, Gypsies, as representatives of exoticism, became fashionable. References to unknown and distant locations, often perceived as completely alien, gave operas, operettas, etc., a trait of distinctiveness, embracing an exoticism that wilfully neglected any actual degree of authenticity and was presented "per se without having to specify the ... exotic land involved" (Dahlhaus 1989: 306). Lacking distinct reference points in the outside world, these exotic places and people seemed to exist outside any fixed historical time. During the performance, the three dimensions of time (past, present, future) collapsed into a 'here and now' that formed a kind of ephemeral 'absolute present' onstage. It can also be argued that these works erased any sense of historical time, as musical exoticism onstage existed as if in no time ("in einem Nu"—to use Heidegger's phrase) (quoted in Gonzalez 2009: 254). In these works, the awareness of the passing of time was denied to non-European peoples, who were treated as if existing outside any historical time, frozen in the unreal, imaginary moment. That type of ahistorical approach to people labelled as exotic was described by Johannes Fabian as the denial of the right to be modern, which he termed *"allochronism"* (1983: 32).

Being part of a broader exotic discourse, Gypsies were frequently depicted on a musical stage. Especially, musical works of the seventeenth, eighteenth, and particularly the nineteenth centuries contributed to the dissemination of stereotypical images of the Gypsy. The Gypsies portrayed in these works tend to appear and suddenly disappear as if mimicking their past appearance in Europe and their travelling customs. Undoubtedly, the introduction of Gypsies on stage provided an additional element of attraction. For instance, in Part VIII of Jean-Baptist Lully's masquerade *Le Carnaval, mascarade mise en musique* of 1668, 'les Bohemiennes' are shown next to Spaniards, Turks, and Egyptians. In the masquerade, they are announced as "une Egyptienne dansante & chantante, est accompagnée de quatre Boëmiennes joüants de la Guittarre" (Lully 1720: 129).

In addition, Gypsies were also featured as central heroes, important for the plot, yet featured on stage for only a short time. The opera *Mignon* (1866) by Ambroise Thomas (based on Goethe's *Wilhelm Meister's Apprenticeship*) can serve as an example of how short and transitory the appearance of Gypsies may be, even though their presence is of crucial importance for the plot. The opera starts with Gypsies being introduced in the opening as dancers, singers, and comedians wandering the world together as a band. The German townsfolk observing the Gypsy dance are clearly aware of the fact that the Gypsies are only passing through their region and accentuate their oriental origins. Among the Gypsies, there is a beautiful girl named Mignon, who, as it turns out, is not really of Gypsy origin but was kidnapped as a small child. The Mignon mentioned in the title grows into a beautiful woman and is forced by the Gypsies to earn her living by dancing and singing so often that it is beyond her capabilities. Gypsies are stereotypically portrayed by the librettists (Michel Carré and Jules Barbier), en masse, as people cruel and brutal towards women. Above all, this short presentation of the Gypsies on stage serves as a mere pretext for introducing some dances.

Similarly, Gypsies appear only for a brief, but crucial, moment in the earlier two-act work *Ideał* (The Ideal) (1840) by Polish composer Stanisław Moniuszko to Oskar Milewski's libretto. The plot concerns the alleged love of a dandy, Karol, for an unknown Gypsy girl called Precjoza or Esmeralda. His affection puts an end to the matrimonial plans that

Karol's father laid out for the young boy. The irony is that the nobleman remains ignorant of the fact that both names belong to fictional Gypsy characters immortalised in literature by Cervantes (Precjoza) and Hugo (Esmeralda). Karol's father and the boy's tutor plan to outwit Karol by pretending that the young man's dreams come true when a group of alleged Gypsies arrive at the court (they are enacted by the servants). Amongst them Karol finds a beautiful 'Gypsy' girl and instantly finds himself enchanted by her. Karol is, of course, unaware of the fact that the girl is in reality Emina, who had earlier pledged herself to him. As in many other similarly conceived works, the appearance of the Gypsies, announced as arrivals from distant Andalusia, constitutes only a pretext for a performance of additional songs and dances. Act Two presents a scene with a Gypsy camp that includes all key components of the romantic vision of a Gypsy world: carts picturesquely situated in a forest, a bonfire smouldering in the centre, the darkness of night heightening the mystery of the situation.

The tendency to introduce Gypsies in light operas, and later also in operettas, as passing heroes was very popular. Their appearance in these works was always justified, yet they would exit the stage as quickly as they had entered it. For instance, the ensemble in Act One, "Steht ein Mädel auf der Puszta," in Georg Jarno's operetta *Die Försterchristl* (The Girl and the Kaiser) (1907), presents the song of the Gypsy Minka, a fortune teller, accompanied by a Gypsy band, praising the beauty of the Hungarian land. The great master of operetta Emmerich Kálmán treated the Gypsy motif in a similar way in his operetta *Gräfin Mariza* (Countess Marica) of 1924. Its action is set in Hungary at the start of the twentieth century. Arriving at her country estate, Countess Marica is greeted by her peasants and Gypsy musicians. Casting the Gypsy woman Mina as the fortune teller clearly points to the stereotypical understanding of Gypsies and their role in society and illustrates how brief and occasional the contact usually was between non-Gypsies and Gypsies.

As already mentioned, the appearance of Gypsies in musical stage works was often (although not always) quite sudden and unheralded. Noted for his ballets, Igor Stravinsky introduced the figure of a Gypsy woman into his *The Fairy's Kiss* (1928), for which he himself created the libretto on the basis of Hans Christian Andersen's *The Ice-Maiden*. His Gypsy woman epitomises mystery, riches (her attire sparkles with valuable jewels), and erotic allure, as her eastern beauty clearly distinguishes her from the rest of the country folk. Obviously, she does not belong to the throng of feting wedding-goers and sticks out because of her exotic accent. Her mysteriousness further manifests itself in her uncertain provenance: she comes from an unknown region and disappears unnoticed, which, again, may be read as an innuendo to the nomadic lifestyle of Gypsies.

Additionally, the association of Gypsies with magicians and sorcerers—mysterious creatures existing outside mundane time—often inspired composers to combine their unknown fate with extraordinary attractiveness, especially with regard to Gypsy women. The custom of introducing mystical figures entering and leaving the musical stage abruptly corresponded well with the widespread knowledge about the wandering style of Gypsy groups. A similar tendency can be also identified in numerous nineteenth-century songs, usually describing Gypsies in a stereotypical manner and drawing on popular clichés. Gypsies, again, were presented as travellers continuously on the move, emerging from the haze of undefined time, and invariably singing their merry and carefree 'tra-la-la,' as if fully immersed in the joy of the moment.

Gypsies, Songs, and Lyrics

In romantic songs Gypsies are often portrayed in carefree choruses, singing their characteristic 'tra-la-la,' as, for example, in the song "Les Bohémiens: Au Bord de la Seine" (1869) by Pompée Barbiano Belgiojoso, based on French lyrics by M. Adolphe Baralle, and devised for bass and baritone with piano accompaniment. The text reproduces the stereotypical notion of the free and highly romanticised Gypsy lifestyle and their connection with nature. It also shows the nonproblematic approach of Gypsies to everydayness: they are travelling without realising the passing of time. Similarly, in Ludovic Benza's (d. 1874) unpretentious song "La Bohémienne," with lyrics by Ali Vial de Sabligny, the refrain portrays carefree Gypsy girls leading a joyful life, blissfully singing 'tra-la-la.' These asemantic words were to become characteristic for songs with Gypsy connotations: in one of Julius Benedict's works (1804–1885), "La Gitane et l'oiseau" for voice, piano, and flute to words by Andréde Badet, 'tra-la-la' is also used in the chorus in alternation with the equally simple 'ah-ah' at the end. Content with the meaningless 'tra-la-la' and enjoying the 'here and now,' Gypsies seem uninterested in anything serious, which is reflected in the lack of 'proper' lyrics in the song. The tendency to present Gypsies in a style bordering on banality continued into the twentieth century, especially in cabaret-like songs of the 1920s and 1930s. In 1930, the English creator of light entertainment songs Lawrence Wright (1888–1964),[3] who wrote under the pseudonym of Horatio Nicholls, composed "Gipsy Melody" to lyrics by Harry Carlton. In this typical Tin Pan Alley short waltz (for voice with a jazz group accompaniment), the 'Gypsiness' marked in the title works as a pretext to reeling off dreams about one's beloved. She appears, *deux ex machina*, together with the sounds of the "Gypsy melody" of the title. In such light songs, Gypsies are shown as people existing outside of time, forever young and jubilant, simply enjoying life. Although they are depicted as travelling, it is never specified which phase of their journey is portrayed—when (and where) they started or when (and where) they are expected to finish it. The aim of their pilgrimage remains unknown. Instead, it is the state of being constantly on the move, which constituted the key characteristics of Gypsies as perceived by non-Gypsy culture.

Although the stereotypical portrayal of Gypsies dominated in the musical culture of the early twentieth century, some contemporary songs featuring Gypsy heroes attempted at exploring additional dimensions of their lifestyle. Although Leoš Janáček's (1854–1928) 'Gypsy' cycle *Zápisník zmizelého* (The Diary of One Who Disappeared) of 1919 also draws on these popular stereotypes, this particular, quite exceptional set of songs occupies a special place amongst the Gypsy songs of the beginning of the twentieth century. In *The Diary of One Who Disappeared* (for voices and piano accompaniment) Janáček (allegedly) used anonymous poems that were initially published in the Sunday editions of the Brno newspaper *Lidové noviny* (People's News) on 14 and 21 May 1916. The paper explained:

> In a mountain village of eastern Moravia, in a mysterious way, there once disappeared J. D., an honest and hardworking young man, his parents' only hope. At the start, thoughts turned to misfortune or crime. Only after a few days was a note found revealing the mystery of the disappearance. This contained a few short verses, which were soon forgotten about, for at the beginning no one considered that these could constitute the key to this mysterious puzzle—as they were seen as simply a notation of folk and country songs. Only a court investigation was to reveal their true content. (quoted in Vogel 1983: 295–296)

---

[3]Wright was also the author of many hits in the style of Tin Pan Alley—including "Dream of Delight" (1916) and "The Toy Drum Major" (1924).

It is later revealed that it was Gypsies who suddenly appeared in the village where the mysterious J.D. (i.e., a boy called Janek) lived. He disappeared soon after their arrival. As suggested in the newspaper, the young lad vanished under mysterious circumstances. No one exactly knew when and how it happened, but during the cycle it is gradually revealed that the youth had fallen in love with a Gypsy girl called Zefka. She became pregnant with his child. Consequently, Janek decided to join the mother of his unborn son and the rest of the Gypsies. Although he knew that his own family would not approve of his decision, he reluctantly accepted the fact that by joining the Gypsies he would be treated as one of them. The dominating sense of fate's inevitability is summed up by Janek in the words "He who goes astray, let him suffer for his deeds" (song xxi), corresponding with the warning that "what is meant for someone cannot be avoided" (song iii and song xviii). The moment Janek chooses to join the Gypsies, his fate is doomed: Once a Gypsy, he is to remain a Gypsy forever. The decision to change his life is not only final but also irreversible.

Musical Language of So-Called Gypsy Compositions and the Principle of Suddenness

In the mid-nineteenth century, instrumental compositions featuring the word *Gypsy* in their titles became very popular (Piotrowska 2013). These works were penned either by professional or amateur composers who tailored them to a salon audience and usually (although not always) were not directly inspired by music performed by the Romany people. Titles such as 'Gypsy dance' or 'Gypsy air' pointed to the imaginary Gypsy world as constructed in literature and other professional musical works. Following common trends of the epoch, compositions alluding to thus-defined 'Gypsiness' were usually associated with Hungarian or Spanish cultures and often named accordingly—as, for example, the so-called Hungarian dances. Although only alluding to 'Gypsiness,' these compositions still managed to recreate the image of Gypsies, which penetrated the minds of European intellectuals.

One of the most characteristic feature of these compositions seems to be their specific treatment of tempo. Structuring time in these 'Gypsy' compositions is subordinated to instantaneous changes and rapid switches, and consequently leads to the fragmentation of the composition. Compositions tend to consist of several sections: they begin with a 'startle' motif and present unprepared, often contrasting ideas, as suggested in the czardas, which is frequently associated with (Hungarian) Gypsies. The omnipresent consciousness of time and its passing results in imbuing time with meaning. In other words, time becomes an independent agency whose flow is regulated by the over-riding principle of 'suddenness' (*Plötzlichkeit*). One can even talk about the rhetoric of musical suddenness in which direct contrast has the crucial role of producing an effect of overall acceleration. But at the same time these prompt changes blur the sense of a stable flow of time on a par with another tool willingly introduced in Gypsy composition, namely, tempo rubato. As a consequence, time as experienced in these compositions seems hazy and distorted: sometimes it is expanded but on other occasions it appears reduced, if not totally abolished. Such speculations on, or playing with, the perception of time—through its rendition, defeat, or a reconciliation with it—simultaneously illustrate a struggling with historical time (i.e., the unproven origins of Gypsies in Europe). These notions also show the eternal aporia of time: its percep-tion may elude our senses, but it can never be totally denied in refer-ence to experiencing music (Green 2011: 62).

The phenomenon of suddenness is not exclusive for music; in fact, it was originally discussed in connection with literature. Suddenness was described, among others, by Aristotle, who dwelt on the concept of *Anagnorisis* (a sudden discovery and realisation of the truth, an unex-pected flash of understanding resulting in a moment of revelation and enlightenment), as closely connected with a sudden change in the action of a tragedy, known as *Peripeteia*. Interpreters of Greek tragedy such as Friedrich Hölderlin or Jacob Bernays saw suddenness as a condition in which one experiences the cathartic effect of tragedy, understanding that violent psychical agitations and upheavals were to be cut by a res-toration of calm. In other words, "drama causes a sudden abreaction of affective energies of the audience that equals cleansing and purification" (Kreienbrock 2013: 16). For Augustine, one of the Church Fathers, sud-denness was linked with Epiphany, an experience of divine interventions resulting in a striking realisation. In the early twentieth century, Martin Heidegger was concerned with the issue of temporality and time, talk-ing about the sudden transition from one to the other, about the sudden leap (Gonzalez 2009: 254). Yet, for Heidegger, "sudden is that type of abrupt which only apparently opposes or contradicts the constant endur-ance" (Davis 2007: 210). This problem of the discontinuity of the self from one moment to another was already discussed in the seventeenth century (Rennie 2005: 116). But it was the Romantic period that privileged the notion of suddenness as an oscillation as well as interference between "presence effects" and "meaning effects" (Gumbrecht 2004: 2), endowing the phenomenon of suddenness with its own aesthetics in which such categories as the unknown, fear, and the demonic prevailed.

If suddenness is a cause of surprise, astonishment, even shock, it is because it stands for something previously unknown and is connected with the pathos of experimentation: consecutive sections of compositions begin as abruptly as sudden surges of contrasting ideas, hence exposing the listeners to the unknowable. Karl Heinz Bohrer argues that "it was the Romantic imagination that first grasped the anticipatory moment—the leap into what just moments before remained unknown—in such a way that a contemporary aesthetics of the unknown as well as our very fear of it could find new life" (1981: 77).[4] Furthermore, it was this state of knowing things that could actually work in a counter-effective manner, as knowledge could be oppressively compelling and lead to a repetition of the past. Thus, it seems possible that knowledge of what had already happened may have been the cause of discomfort because it painfully signalled the irreversibility of the past and the impossibility to reclaim it: it also indicated the subservience to history (Rennie 2005: 204).

But while the past remains immutable—it at least appears to be stable—the future seems flexible to the extent that it can be influenced, if not completely created. The acute awareness of the existence of the future, its mystery and unknown status, might cause certain apprehensions, and even fear, especially if fate is about to come as a surprise, leaving a person unprepared. Søren Kierkegaard saw suddenness as a central category in his theology of demonic despair, claiming in 1844 that anything demonic is of rapid nature. The role of fate in determining a future that always comes unexpected, even if foreseen, is central to Georges Bizet's opera *Carmen* (1875), which is perceived as the quintessential 'Gypsy' work in European culture. The destiny of the main protagonist, her sudden death, although foretold by her friends Frasquita and Mercedes reading cards, still seems incomprehensible: too abrupt, too violent, too tragic.

---

[4] Translated in Rennie (**2005**: 116).

Suddenness seems satanic, similar to Gypsies who have always been regarded as demonic. Especially their dark complexion encouraged their demonisation, for in early Christian Europe black colour was already linked with impure powers and even to the devil himself (di Nola 2004: 311–312). Gypsies, for centuries demonised as a result of their skin colour, Eastern origins, and cultivation of their own customs, constituted within Europe a marginalised group. They were considered to be the bearers of a certain "ceremonial or mythological legacy which is dangerously opposing those models suitable for the majority, becoming sorcery by dint of its antithetic relation to these models" (ibid. 231 [my translation]). The association of 'Gypsiness' with suddenness prompts, in works dedicated to them, an association with a leap into the unknown; 'Gypsiness' is even likened to anarchy, that is, to the breaking of rules, be it with regard to the harmonic or regulating tempos of compositions. Hence, the principle of suddenness observed in 'Gypsy' compositions ties into and corresponds well with the still undiscovered Gypsy world, bringing to the fore such feelings as terror and lack of trust (Bohrer 2005: 58). And, as possibly one of the strongest arguments for claiming authenticity is a quotation (ibid., 28), many of these 'Gypsy' compositions often contained tunes that had originated on Hungarian soil and were considered as Gypsy. These tunes were often cited, more or less verbatim, to validate the authentic 'Gypsiness' of these compositions.

On the structural level, suddenness affects the coherency of the composition and its temporal structure, as it "is not merely a formal result of narrative technique" (ibid., 55) but of its imposed internal logic, turning the piece of music into a fragmentary narrative that creates the appropriate context for various musical gestures and figures (ibid., 20). The sudden historical appearance of Gypsies is celebrated in these compositions anew, underlining the importance of the particular presence and 'presentness' of their being. It is this notion of presence, of experiencing an actual musical moment, that fascinates the listeners of these compositions (De Man 1971: 152). The principle of suddenness excludes anticipation, promoting, in its stead, the intuitive premonition that something will change, that something will happen during the composition. The epiphany of the moment works as an allegory of the present and as a negation of the concept of undisturbed continuity. Instead, we experience sudden engagements and emphatic feelings, which serve as harbingers of both action and (rapid) change. Furthermore, suddenness often suggests the appearance of something new, almost as if suddenness worked as a guarantee of innovative and fresh ideas (Bohrer 2005: 20). Suddenness ascribes itself into the stylistics of 'never-before-seen' and 'unheard-of' adventure. When some new musical idea is abruptly introduced, for example, when a melodic motif appears that belongs neither to the past (anymore) nor to the future (at least not yet), the importance of the moment is highlighted by this very suddenness, which points to the reduction of time to a particular point in time (Heidegger 1996: 373–374). Although suddenness is characterised by an abrupt, quick start, it is not however restricted to any specific duration, nor is it necessarily executed quickly, in a hasty manner.

The principle of a sudden incident observed in these compositions suggests that the introduction of changes, although following a certain pre-compositional, pre-planned concept, may seem of accidental character. Hence, the suddenness of these changes entails bewilderment caused by sentimental vagaries of whimsical alterations, or even bafflement instigated by the introduction of poignantly bombastic moments contrasted with blithe passages.

However, the sheer suddenness, that is, abruptness, of introducing musical ideas cannot solely be considered the most characteristic feature of compositions titled 'Gypsies.' It is rather an amalgam of various elements, including overexposure of augmented seconds skilfully played out in minor tonalities and an abundance of rich ornamentations. Yet, it is the principle of suddenness that remains a very characteristic feature of these compositions, although it may, obviously, also appear in other musical works deemed to be exotic and referring to far-away destinations in the titular or textual layer. It seems that playing with structures of time is one of the most prominent feature of compositions called or associated in the nineteenth and twentieth centuries with 'Gypsiness.' Compositions that announce the presence of the 'Gypsy' in their titles and adhere to the principle of suddenness usually contain fragmented structures reminiscent of mosaic or montage of shorter sections. Some of them introduce the form of czardas as a dance willingly associated with Hungarian Gypsies, and accordingly usually consist of two opposite parts. However, even this simple dichotomous form based on the contrast of tempos (slow in *lassan*, fast in *friska*) is fragmented, allowing for an element of unpredictability as well as for a feeling of disjunction, because the appearance of opposing sections always comes as a surprise: it is unexpected.

For example, the famous Hungarian Rhapsodies by the renowned Romantic composer Franz Liszt follow the simple arrangement of the czardas, preserving even the original labelling by dividing rhapsodies internally into *lassan* and *friska*. The composer also made use of dance melodies derived chiefly from the traditional *verbunkos* and *czardases* (or at least they are passed off as folk music or were the original works of nineteenth-century Hungarian composers). The temporal structure of these compositions attests to their 'Gypsy' associations. At the same time the Hungarian Rhapsodies are representative of the *style hongrois* (Bellman 1993: 196) and abound with harmonic freedom: They contain numerous unexpected enharmonic modulations and spontaneous semi-tonic shifts (Malvinni 2004: 105), which may also be associated with the suddenness and abruptness ascribed to Gypsies.

Also, the Hungarian Dances by Johannes Brahms are in fact *czardases*. Here the harmonious and melodic solutions as well as the stimulation of sounds typical of a Gypsy orchestra create a certain *Stimmung*. The alleged Gypsy dimension of the composition manifests itself in the use of themes of varied provenance, sometimes already exploited by Liszt but chiefly by Hungarian composers. Brahms rarely cites a melody in its entirety. Instead, he chooses shorter fragments, combining and modifying them to create a sense of authenticity and an impression of a refined and fantastic Gypsy world. In terms of form, the *Hungarian Dances* duplicate the ternary scheme, with recourse, however, to the czardas through employing contrasting tempos.

In the second half of the nineteenth century, *czardases* became especially popular, performed as salon miniatures on the piano or violin. In *Tzigan: Impromptu pour piano* (1887) by Marie de Verginy, for example, the contrast in tempos has an important role, as does the effect of imitating the sound of the dulcimer (arpeggio chords, the characteristic repetitions of two pitches in a very quick tempo). Through rhythm, the composer introduced numerous syncopations, grace notes as well as tuplets, allowing for the possibility of employing a rubato tempo. In his *Zigeunerweisen*, opus 20, for violin and piano, also for violin and orchestra (1878), Pablo Sarasate, utilising the scheme of the czardas, followed the same style, adding to the popularity of salon miniatures conceived as *czardases*, which lasted until the early twentieth century. Gustave Michels also composed *Bohêma Czardas* for piano in 1908, but perhaps the most famous czardas at the time was Vittorio Monti's *Csárdás* (1904), which consists of a number of contrasting sections. Monti's *Zingaresca* for violin with piano accompaniment (1912) is also marked by repetitions of the same musical material and features numerous changes in key and tempo. Similarly, at the beginning of the twentieth century, Fritz Seitz (1848–1918) composed *Zigeuner Kommen*, opus (op.) 16, no. 4, for violin with piano. The piece represents an exemplary type of miniature devised as a czardas, preserving the latter's characteristic format. Also "Gipsy Fantasy" for violin and piano (1938) by Dave Rubinoff directly links to the czardas, not only through the contrast in tempos but also through its arrangements (cantabile *lassan*, figurative *friska*).

In the early twentieth century, Maurice Ravel's orchestral poem *Tzigane* (1924), initially written for violin and piano and then for violin and orchestra, connected to Gypsy music in its Hungarian guise through the excitation of an augmented second interval. Furthermore, it was reminiscent of the Gypsy tradition of violin playing and included considerable technical demands on the performer of the solo violin part. Full of virtuosic passages such as cadenza abounding in flageolets, double stoppings, and pizzicato, *Tzigane* not only provides an opportunity for violinists to show off their exquisite technique but also enables the musician to display the whole palette of moods, being, according to Jonathan Bellman, a "recreation of the Gypsy mood and language filtered through Ravel's own impressionistic sensibility" (1993: 217). Moreover, with its frequent changes in tempo, the work is reminiscent of the czardas. Superficially a rapturous piece with a medley form, *Tzigane* has a very well organised inner structure with a division into three interconnected parts: (1) an initial violin solo cadenza, (2) the presentation of a theme with driving rhythms followed by dazzling variations, and (3) the type of finale that starts out slowly only to speed up the tempo to conclude the work in the manner typical of czardas, to which this work obviously alludes. After the cadenza, we are offered the slow part—*lassan*—and then the fast part—*friska*. Additional alterations of the tempo (alternately suggesting ritardando and accelerando) magnify the overall impression of the composition as some type of jotted-down improvisation (by a Gypsy violinist?), just as the tonality of the work offers breaks from traditional harmony in the form of unexpected solutions such as progressions of somehow 'distorted' chords in the last section, providing supplementary "auditory benefits to the listener" (Manns 2007: 116–117). These effects can be interpreted as suggesting the unorthodox approach to music-making as observed among self-taught Gypsy musicians or be perceived as an attempt at emulating the out-of-tune sound of the cimbalom played by the busking Gypsies.

In all 'Gypsy' compositions, setting together sections that may tessellate, echo, or even foreshadow each other, juxtaposing tempos, rhythms as well as tonalities and textures can lead to the effect of an overt sense of disruption. Changes occurring without warning and unprepared combinations of different motifs may result in the spurious lack of linear continuity. Moreover, the internal progression of these works is secured by the impulsive appearance of short sections and brief moments of rapid leaps that constitute their inner force of development. The feeling of Gypsy spontaneity is pursued by introducing certain solutions mimicking hesitation and hastiness: the unexpected outbursts of ideals are balanced with more focused, contemplative fragments. Contrasts in various 'Gypsy' compositions are evoked, for example, by the introduction of divergent themes in different tonalities or modifications of the mode (major–minor as in *Zigeuner-Liebe und Leben: Rhapsodie Zingane* [1905] for violin and piano by Franz Carl Bornschein (1879–1948) or *Hungary* [*Rapsodie Mignonne*], op. 410, [1907] for piano by Carl Kölling [1831–1914]). The effect of surprise is asserted by unexpected changes in the direction of the melodic line, juxtaposition of conflicting metric groupings reversing the character of the presented tunes, and opposing tempos of neighbouring sections (slow–fast), etc. Such huge contrasts generate the sense of dramatic narration supported by the volatile order of the presented fragments, creating additional energy. Indeed, so-called 'Gypsy' compositions—rich in diversity and abundant in ideas—seem to be obsessed with the rapid proliferation of musical motifs stimulating various emotional responses.

The suddenness of such changes causes alternating emotional responses in the listeners. These rapid changes of emotions that are elicited can be related to the stereotypical image of Gypsies who were accused of a childlike changeability of moods. Hence even the compositions that pointed to the 'Gypsy' in their titles often strengthened that popular belief by including phrases such as capriccio or fantasy. Furthermore, the suddenness of change is coupled with their intense accumulation: sometimes alterations appear as often as every 32, 16, or even 8 (!) bars, as in *Zigeuner-Fantasia* for piano (1905) by Gyula von Bartay. In this piece the composer introduced new tempos, new metres, and new tonalities no less than 30 times. The montage-like structure (within overall ternary form) is also characteristic of *Zingara: Fantaisie* (1894) by Regina Beretta. Sometimes the composition seems to toy around with time (and to point to alternative time or times), as in C. Reichenbach's *Tzigana*. For this piece, the composer suggested slowing the tempo every few bars, thereby imitating the emotional manner with which Gypsy musicians play.

As if imitating free Gypsy spirit, in many 'Gypsy' compositions composers followed a loose-form organisation, creating their works as a hotchpotch of various melodic, harmonic, or rhythmic ideas. David Malvinni commented on the idea of montage, mosaic, or medley, that is, of putting together different, if not contrasting, patches and episodes, and especially their sudden and unprepared introduction in 'Gypsy' compositions, and suggested using the concept of potpourri in these cases: "Pasting themes together, cutting them up, without any musical logic, without *Geist*, the absence of work ... escaping the purview of musicology are also the ironic features of musical language or 'dialect' that most easily speak to and aurally 'touch' audiences" (2004: 84). With regard to compositions conceived as 'Gypsies,' the idea of constructing their narrative by joining different musical ideas in an unexpected way was crucial. Can the nature of these compositions be described as disrupted or disconnected? I would argue that, in contrast to what might be speculated, the suddenness, although prone to result in fragmentation, is responsible for the ramification of constructing the perceived narrative and also assures the sense of inward direction, working as a guarantee of the onward—often kinetically irresistible—movement. Hence this type of construction enables the perception of sections as introduced suddenly yet logically, as this 'moment-by-moment' narrative secures the credibility of the composition, its coherency, and prompts listeners to engage with the fragmentation as an end in itself.

WHY 'GYPSY' COMPOSITIONS BECAME SO POPULAR

Although 'Gypsy' compositions could be regarded as symbolic of the Romantic period, during the Biedermeier period they soon took on the role of a substitute and became a platform of artificially invoked emotions, which, once the performance was over, could be easily forgotten or suspended until the next performance. The term *Biedermeier* as I use it in this chapter does not so much relate to a historical period but instead refers to a specific type of sensitivity as well as certain moods, tendencies, and feelings that were reinforced in a period of relative peace and stability. Although reinforcing security and strengthening traditional values such as piety or simplicity, the epoch also strongly promoted conventionality, putting emphasis on such categories as nostalgia and harmony. Characteristic aesthetic standards connected with Biedermeier include sentimentalism, affection, sensitivity, gentleness, moderation, and modesty, as well as *Gemütlichkeit*, understood as a state of cosiness, warmth, and friendliness. In such an environment the conflict between ideals and reality was acutely felt (Kubiak 2006: 18), as it fostered the awareness of the greyness and gloominess of human life. Above all, it helped realise to what degree an inhabitant of a city is subordinated to his or her destiny and—unable to effectively change this situation—bound to resign. However, giving up sentimental inclinations was considered a bourgeois (*bürgerlich*) virtue, because it was believed that a man should observe the ideal by serving reality. Following the footsteps of great Romantic heroes experiencing brave adventures did not constitute the ultimate aim for petty bourgeoisie, but they did not completely forget about the pathos of the soul's turmoil, about spiritual needs, sorrows, anguish, and other extreme emotions. Rather, these were sublimated into the world of phantasies opposed to rationality: dreams were confronted with intellect, desires calmed down by reason.

With regard to music, Biedermeier people could experience these thrilling emotions without the danger of being exposed as rebelling against the dominating tendency. Furthermore, as I argue, the 'Gypsy' theme was used as a pretext to introduce extremely polarised feelings into compositions, thus allowing listeners to enact these feelings within their own private sphere, while at the very same time keeping up the appearances of proper citizens, respected fathers and mothers, and honourable clerks and officials. In the period of Biedermeier, when small social gatherings over a cup of tea or coffee were cultivated (Herman 2006: 70), the acclaimed bourgeois lifestyle continued, allowing for a common experience of listening, playing, and enjoying musical compositions. The choice of the repertoire seems symptomatic, as short forms such as salon miniatures for the piano were preferred. Often these forms were easy to perform, even without professional musical training. Numerous publications for in-home music making were published, encouraging singing (e.g., German lieder) or playing together (e.g., arrangements of operatic excerpts from the most famous works of the time). The broadened reach of music facilitated the enactment of intense emotions. In the eliciting of passions and forceful emotions through music, suddenness and the epiphany of every single moment played a significant role. Suddenness, which was so characteristic of 'Gypsy' compositions, did not serve as a kind of esoteric *chiffre* or as an allegory of the present or of the negation of continuity. Instead, Biedermeier people, who were denied a transgression of tradition in real life, were prone to listen to compositions that followed the principle of suddenness. The more they were aware of the impassivity and the torpor of their own heavily regulated lives, the more they longed for something spontaneous and impetuous, perhaps even perfervid and impellent.

In this context, the 'Gypsiness' announced in the title of a composition must have seemed like an invitation to enter the world of passions but also the freedom that Gypsies enjoyed. In other words, 'Gypsy' compositions, especially those following the principle of suddenness, offered the possibility of engaging in great emotions in a quite safe mode, that is, in the cosiness of the salon's armchair. The transfer of emotions into this new, domesticated dimension allowed for specific sensations such as experiencing different levels of time and space, including travelling in time, while simultaneously being and not-being the subject of intense emotions. More precisely, the 'Gypsy' compositions worked as stimuli or, rather, as surrogates of 'real' emotions experienced in 'real' life. These works would warrant a substitution of strong emotions by focusing on the abruptness of the moment, which acquired its spontaneous, emotion-related meaning immediately on the spot, not prospectively as a calculation of its reference to the future. The rhetoric of these moments characterised by suddenness produced emotional intensity, which, as I argue, substituted the passions and upheavals of the outside world that people were (most probably) trying to shun in their real life. But it is never possible to escape these feelings entirely, at least not in the sphere of dreams and longings, because these types of emotions are raised, for example, while reading about handsome heroes of romantic adventures, or watching pictures. Sentiments thus evoked are then forever longed for, missed, and consequently recreated over and over again in an artificial way that makes use of the principle of suddenness as well as of the acute feeling of the present, be it in the form of bungee-jumping in the contemporary world or listening to compositions bearing 'Gypsy' titles.[5] Just as bungee-jumping is predominantly concerned with the 'here and now' and focused on the dreaded moment about to arrive, so are 'Gypsy'

---

[5] In contemporary America, the role of raising surrogate emotions is performed, for example, by live transmissions of police car chases. One of the most spectacular cases took place in 1994 when 95 million people followed a live broadcast of the police chasing after footballer O.J. Simpson, accused of murdering his wife. See Furno-Lamude (1999: 23).

compositions, even if they are working on a different level, referring pre-dominantly to other senses. What remains common for both, as a result of the essence of human nature, is the obsession with the moment that approaches suddenly, without warning, and never persists.

One would be tempted to comment: just like the Gypsies.

BIBLIOGRAPHY

Bellman, Jonathan. 1993. *The Style Hongrois in the Music of Western Europe.* Boston: Northeastern University Press.

Bohrer, Karl Heinz. 2005. *Nagłość,* trans. Krystyna Krzemieniowa. Warszawa: Oficyna Naukowa.

———. 1989. *Nineteenth-Century Music,* trans. J. Bradford Robinson. Berkeley: University of California Press.

Davis, Bret W. 2007. *Heidegger and the Will: On the Way to Gelassenheit.* Northwestern University Studies in Phenomneology & Existential Philosophy. Evanston: Northwestern University Press.

De Man, Paul. 1971. *Blindness and Insight: Essays in the Rhetoric of Contemporary Criticism.* New York: Oxford University Press.

di Nola, Alfonso M. 2004. *Diabeł,* trans. Ireneusz Kania. Kraków: Universitas.

Fabian, Johannes. 1983. *Time and the Other: How Anthropology Makes its Object.* New York: Columbia University Press.

Fraser, Angus. 2007. *The Gypsies.* Malden, MA: Blackwell.

Furno-Lamude, Diane. 1999. The Media Spectacle and the O. J. Simpson Case. In *The O. J. Simpson Trials: Rhetoric, Media, and the Law,* ed. Janice E. Schuetz and Lin S. Lilley, 19–35. Carbondale: Southern Illinois University Press.

Gonzalez, Francisco J. 2009. *Plato and Heidegger: A Question of Dialogue.* University Park: Pennsylvania State University Press.

Green, Edward. 2011. Steiner, Korngold and the Musical Expression of Physical Space: A Preliminary Note. *International Review of the Aesthetics and Sociology of Music* 42 (1): 59–78.

Gumbrecht, Hans Ulrich. 2004. *Production of Presence: What Meaning Cannot Convey.* Palo Alto: Stanford University Press.

Heidegger, Martin. 1996. *Being and Time,* trans. Joan Stambaugh. Albany: State University of New York Press.

Herman, Georg. 2006. Człowiek Biedermeieru, trans. Łukasz Musiał. In *Spory o Biedermeier,* ed. Jacek Kubiak, 67–71. Poznań: Wydawnictwo Poznańskie.

Kenrick, Donald. 2010. *The A to Z of the Gypsies (Romanies).* Lanham: Scarecrow Press.

Kreienbrock, Jorg. 2013. *Malicious Objects, Anger Management, and the Question of Modern Literature*. New York: Fordham University Press.

Kubiak, Jack. 2006. Introduction to *Spory o Biedermeier*, ed. Jacek Kubiak, 7–63. Poznań: Wydawnictwo Poznańskie.

Lully, Jean-Baptiste. 1720. *Le carnaval, mascarade mise en musique*. Paris: De l'Imprimerie de J.B. Christophe Ballard.

Malvinni, David. 2004. *The Gypsy Caravan: From Real Roma to Imaginary Gypsies in Western Music and Film*. New York: Routledge.

Manns, James W. 2007. The Concept of Unity in Music. In *What Kind of Theory Is Music Theory? Epistemological Exercises in Music Theory and Analysis*, ed. Per F. Broman and Nora A. Engebretsen, 107–131. Stockholm: Stockholm University.

Piotrowska, Anna G. 2013. *Gypsy Music in the European Culture: From the Late Eighteenth to the Early Twentieth Centuries*. Boston: Northeastern University Press.

Rennie, Nicholas. 2005. *Speculating on the Moment: The Poetics of Time and Recurrence in Goethe*. Göttingen: Wallstein.

Rochow, Ilse, and Klaus-Peter Matschke. 1991. Neues zu den Zigeunern im byzantinischen Reich um die Wende vom 13. zum 14. Jahrhundert. *Jahrbuch der Österreichischen Byzantinistik* 41: 241–254.

Vogel, Jaroslav. 1983. *Janaczek*, trans. Henryk Szwedo. Kraków: Polskie Wydawnictwo Muzyczne.

CHAPTER 10

The Present: An 'Unknown Time' in the German *Kaiserreich* around 1900

Caroline Rothauge

In his essay *Berlin: Ein Stadtschicksal* (Berlin: A Town's Fate), first published in 1910, Karl Scheffler, a German art critic, lamented that in Berlin main meals were not being served at a specific time anymore:

> One does not leave business until very late. ... All of the attempts to introduce something like the English working time in Berlin and thus to give business life as well as family life a certain form have failed so far. One cannot even get the masses to reach such practical unanimity. One consequence of this is that there are no fixed hours for main meals in Berlin. One finds circles of citizens where lunch is taken in a provincially solid manner at 1 or 2 o'clock [p.m.], a snack at 4 o'clock, and dinner around 8 o'clock; in other sections of the middle class one takes lunch not before 3 or 4 o'clock and one speaks of 5 o'clock tea. It always depends on business, on social status, and on concepts of refinement. ... In any case, a single hour uniting all of the city's population for the main meals, which accords such characteristic traits to city life like in Paris or London, does not exist in Berlin. Eating habits are as good as ruled by anarchy. ...

C. Rothauge (✉)
Lehrstuhl für Neuere und Neueste Geschichte, Katholische Universität
Eichstätt-Ingolstadt, Eichstätt, Germany

© The Author(s) 2017
S. Baumbach et al. (eds.), *The Fascination with Unknown Time*,
DOI 10.1007/978-3-319-66438-5_10

211

A strange atmosphere prevails in this unregulated city life. One person is still hungry, whereas the other is already sated; one drinks his afternoon coffee next to people having breakfast and when those are taking their afternoon coffee there are paying guests who are already having dinner before theatre. With such a disorder, mutually shared social events cannot come about. (Scheffler 1910: 212ff. [my translation])[1]

A number of different observations, opinions, and fears become evident here: Not only did people work long hours, but professional and private lifetimes were converging. According to Scheffler, no longer was there any recognisable structure or order regarding mealtimes. The diverse uses of time he observed differed according to individual or group-specific preferences. Therefore, Scheffler stressed two points: First of all, the asynchronicity of daily modern city life increased. Second, the disposition and possibilities for simultaneous and shared experiences decreased. Clearly, Scheffler interpreted the multiplication of different ways to structure time in modern city life as a fragmentation that ultimately facilitated social disintegration.

In this context, it is not important whether Scheffler's judgement is right or wrong. However, it serves as a significant example of the perception of one intellectual to whom the present became increasingly confusing and thus unknowable in the capital of the German *Kaiserreich* at the beginning of the twentieth century. This essay discusses unknown

---

[1] "Denn man kommt spät erst aus dem Geschäft. ... Alle Versuche, in Berlin etwas wie die englische Arbeitszeit einzuführen und dadurch dem Geschäftsleben sowohl dem Familienleben eine bestimmte Form zu geben, sind bisher gescheitert. Selbst zu solcher praktischen Einmütigkeit sind die Massen nicht zu bringen. Eine Folge davon ist, daß es für die Hauptmahlzeiten in Berlin keine festen Stunden gibt. Man findet Bürgerkreise, in denen provinzlich solide um 1 oder 2 Uhr zu Mittag gegessen, um 4 Uhr gevespert und gegen 8 Uhr zu Abend gegessen wird; in anderen Teilen des Mittelstandes ißt man erst um 3 oder 4 Uhr zu Mittag und spricht vom Fünf-Uhr-Tee. Es richtet sich immer nach dem Geschäft, nach der sozialen Stellung und nach den Begriffen von Vornehmheit. ... Eine einzige Stunde, die die ganze Stadtbevölkerung zu den Hauptmahlzeiten vereinigt und dem Leben der Stadt so charakteristische Züge verleiht wie in Paris oder London, gibt es in Berlin jedenfalls nicht. Es herrscht in den Speisegewohnheiten fast Anarchie. ... Es ist eine seltsame Stimmung in diesem ungeregelten Stadtleben. Der Eine hat noch Hunger, während der Andere schon gesättigt ist, man trinkt seinen Nachmittagskaffee neben Frühstückenden; und wenn diese ihren Nachmittagskaffee nehmen, sitzen Gäste da, die vor dem Theater schon zu Abend speisen. Bei solchem Durcheinander kann es zu gemeinsamen gesellschaftlichen Veranstaltungen nicht kommen."

experiences of time in the German Empire around 1900. It argues that the "component of the unknown" (Koselleck 2000: 164 [my translation])[2] in contemporary time experiences was not acceleration but rather the yet again increased pluralisation of both the notions and uses of present time(s). Experiencing pluritemporality provoked a series of expressions of opinion, all of them combining repudiation with fascination. In that sense, commenting unknown modes of present time around 1900 in *Kaiserreich* Germany was highly ambivalent and as such typical of modernity.[3]

MODERN TIMES: NOTHING BUT ACCELERATED?

Velocity and *speed* are the buzzwords most commonly associated with a specifically modern experience of time. The thesis that modernity distinguishes itself by means of a specific understanding of time is a rather common one (Jung 2010–2011: 172–174; Stockhorst 2006: 157). It has been labelled as the "model of temporalisation of modernity" (Jung 2010–2011: 172 [my translation])[4] and, as such, has obtained the status of a master narrative. The historian Reinhart Koselleck focused a large part of his research on the relationship between modernity and time and helped establish said master narrative. His argument that the specifically modern way to relate to time is 'acceleration' has been widely adopted in German humanities discourse.

The origins of Koselleck's preoccupation with a theory of modernity lay in the so-called history of concepts (cf. Koselleck 1972: xiii–xxvii),[5] a sub-category of German social history that Koselleck himself and a few other historians developed in the late 1960s and early 1970s. In the course of his research, Koselleck elaborated on the observation that classical topoi radically changed their meaning during a transitional period from 1750 to 1850.[6] According to Koselleck, this change resulted from the necessity to adapt to the conditions of the modern world (cf. ibid.),

---

[2] "Unbekanntheitskomponente".

[3] Ambivalence is considered to be intrinsic of modernity, cf. Middendorf (2012: 148).

[4] "Verzeitlichungsmodell der Moderne".

[5] "Begriffsgeschichte".

[6] For this transitional period, Koselleck coined the term "Sattelzeit" (Koselleck 1972: XV).

one of these conditions being a radically new experience of time, an experience of time in which acceleration plays a key role.[7]

To Koselleck, acceleration refers to a change in the way time was felt and in the way one was becoming aware of time at the turn of the eighteenth to the nineteenth century (Koselleck 2000: 163). This change was characterised by the fact that "everything changes faster than one could have expected or anyone had experienced up to this point. Due to shorter timespans, a component of the unknown enters daily life …" (ibid., 164 [my translation]).[8] The two concepts "space of experience" and "horizon of expectation" are essential for Koselleck's explanation of the new experience of acceleration (Koselleck 2004: 255–275). According to Koselleck, both concepts intertwine past and future and are therefore suitable to broach historical time (Koselleck 1979: 353; Koselleck 2004: 262). Koselleck focussed on the turn of the eighteenth to the nineteenth century, because it marked a point in history where "expectations have distanced themselves evermore from all previous experience" (Koselleck 2004: 263). This asymmetry between the "space of experience" and the "horizon of expectation" led to the emergence of a "new time" that was distinctively different from all former historical times. The first concept that had been used to grasp this asymmetry was 'progress' (ibid., 268). Koselleck, however, pleaded for the concept of 'acceleration' instead, because it "involves a category of historical cognition which is likely to supersede the idea of progress conceived simply in terms of an optimization …" (ibid., 270).

To Koselleck, acceleration distinguishes the experience of time since the turn of the eighteenth to the nineteenth century from all experiences of time people had had before. Indeed, acceleration has been observed and complained about repeatedly in Western countries from

---

[7] In response to this line of argument, some of Koselleck's colleagues distil a theory of historical time from his approach to a "history of concepts" (Dipper 2000: 308). All in all, the issue of time and people's perception of it in different historical contexts has hardly been investigated in German Contemporary History, cf. Geppert and Kössler (2015: 10–11), Graf (2012), Schwarz (2001: 451). In Germany today, it is mainly sociologists who stress that acceleration is specific to modernity, with Hartmut Rosa meeting with the biggest response in the feuilleton, cf. Rosa (2005), Rosa (2013).

[8] "sich … alles schneller ändert, als man bisher erwarten konnte oder früher erfahren hatte. Es kommt durch die kürzeren Zeitspannen eine Unbekanntheitskomponente in den Alltag …".

the mid-eighteenth century on (Dipper 2000: 303; Schwarz 2001: 468). One well-known quote in this context is taken from the autobiography of the US-American historian Henry Adams and refers to what he labelled as the "law of acceleration":

> The movement from unity into multiplicity, between 1200 and 1900, was unbroken in sequence, and rapid in acceleration. Prolonged one generation longer, it would require a new social mind. As though thought were common salt in indefinite solution it must enter a new phase subject to new laws. Thus far, since five or ten thousand years, the mind had successfully reacted, and nothing yet proved that it would fail to react—but it would need to jump." (Adams [1907] 2011: Chap. xxxiv)[9]

Taking into account the consequences of a supposed acceleration, Adams still described the development between 1200 and 1900 as a "movement from unity into multiplicity" and hence as a movement towards plurality. This observation should not be ignored as it points to the following question: What other perceptions of time have existed besides acceleration? Acceleration might have been an initially new and central experience of time, but it was not the only one, nor unanimous, let alone uncontested.

Complaints uttered by some experts in early modern history point to a similar direction: They have criticised superficial readings of Koselleck's texts, which suggest that the experience of time in modernity was and is a uniform one (Jung 2010–2011: 179). In this context, Koselleck's model of temporalisation of modernity can be interpreted as a theoretical framework that supports a teleological narrative of progress and thus modernisation theories. Such a categorical and generalising understanding of the relationship between modernity and time neglects, as might be argued, "the highly diverse currents within the discourse of time" (Stockhorst 2006: 158 [my translation]).[10] If one wants to hold on to the idea that modern life is characterised by a specific experience of time, the common denominator can no longer be a general experience of acceleration but should rather be "the confrontation with the increasing

---

[9] Adam's autobiography, entitled *The Education of Henry Adams*, was first published in 1907 and was awarded a Pulitzer Price posthumously in 1919.

[10] "die höchst unterschiedlichen Strömungen im Zeitdiskurs".

incommensurability of different time cultures" (Jung 2010–2011: 179 [my translation]).[11]

Koselleck himself repeatedly alluded to the "contemporaneity of the noncontemporaneous" (e.g., Koselleck 2004: 239, 246). He also stressed that in the "new time," "the one process of time became a dynamic of a coexisting plurality of times," because "generations did live in the same experiential space, but their perspective was interrupted according to political generation and social standpoint" (ibid., 269). Moreover, Koselleck emphasised that time itself is subject to historical change, depending on the connection between the "space of experience" and the "horizon of expectation" in a specific historical context. Bearing that in mind, Koselleck's model of temporalisation of modernity as such should not be rejected altogether. Rather, the task is to look at the relationship between modernity and time in a more differentiated way and to modify the model correspondingly. This challenge applies even more to the industrialised countries of the Western Hemisphere around 1900, where acceleration was not something that had never been previously experienced. Hence, at this stage, acceleration cannot possibly have been an "unknown component" in people's daily lives and the ways in which they experienced time. Following Koselleck, more than 100 years had passed since the emergence of the modern world. Thus, the experience of acceleration would have widely matched the expectation of acceleration around 1900. Instead, the unknown in notions and uses of time was rooted in the yet again increased pluralisation of what people experienced and constructed as to be 'the present'—or, rather: 'their present.'

EXPERIENCING PRESENT TIME(S): SYNCHRONISATION VERSUS PLURALISATION

In contemporary history today, it is commonly assumed that another "cultural threshold" (Dipper 2010; Seefried 2015: 37 [my translation])[12] was crossed in the Western industrialised countries around 1880. In Germany, processes of mechanisation, scientification, bureaucratisation, democratisation, and urbanisation intensified in the late *Kaiserreich*,

---

[11] "die Konfrontation mit der zunehmenden Inkommensurabilität unterschiedlicher Zeitkulturen".

[12] "Kulturschwelle".

especially in the years around 1900 (Kroll 2013: 9; Hübinger 2014: 1, 8–9, 20–21; Seefried 2015: 37). These tendencies and developments suggest that sociopolitical structures as well as daily living environments were again changing. Consequently, new patterns of order were required[13] that met the needs of the new epoch and included ways of contending with time. However, what historians actually consider crucial for entering this new era, optionally called "classical modernity"[14] or "high modernity,"[15] relates to people's self-perception at the time. People living in the Western industrialised countries increasingly perceived the historical period in which they lived as 'modern' from the 1880s onwards (Dipper 2010).[16] This widespread mode of self-reflection also encompassed considerations about experiences of time in modern life.

Although a "strikingly simultaneous emergence of 'time talk'" (Ogle 2015: 9; cf. Ogle 2013: 1377, 1383) can be observed around the globe in the second half of the nineteenth century, in the German *Kaiserreich* it was especially the turn from the nineteenth to the twentieth century that provoked an increase in statements and analyses about how to evaluate past, present, and future. Contemporaries perceived 1900 as a caesura, as a "big turning point in time" (H. Landsberger, quoted in Puschner 2010: 536 [my translation]).[17] In newspaper articles that covered the turn of the century, one type of self-characterisation was especially frequent: People referred to themselves as "people of transition", believing "the present [to be] in a big process of transformation" (*Neue Deutsche Rundschau* 11 (1900); quoted in Puschner 2010: 532

---

[13] For further comments on those new patterns of order required after 1880 with far-reaching consequences for the twentieth century, see Raphael (2008: 85–91).

[14] This term is attributed to the German historian Detlev K. Peukert (1987).

[15] A labelling that can be traced back to the German historian Ulrich Herbert (2007).

[16] According to this, modernity was and is understood as a historical era on the one hand but also as a diagnosis of contemporarily present time(s) on the other. The latter understanding of modernity implies a specific narrative perspective, as the one apparent in modernisation theories, for example, in Dipper (2010). Similarly, the literary theorist Hans Ulrich Gumbrecht differentiates between a contemporary understanding of modern phenomena as a current state which is experienced as an era and a perspective which focussed on modernity as a process, namely as the past of a future present, see Gumbrecht (1978: 96, 120ff). A third understanding of modernity defines it more narrowly as a stylistic movement in literature, music, art, or architecture, see Dipper (2010).

[17] "große[n] Zeitwende". Cf. Kroll (2013: 10).

[my translation]).[18] Apparently, they were especially intrigued by the temporal dimension of the present.[19]

The overall perception in Germany around 1900 was that of living in a highly dynamic time, a time of upheaval where fixed parameters now barely existed (Puschner 2010: 533; Salewski 1971: 354). At first glance, these broadly published convictions seem to contradict widespread ideas on the nineteenth century as a period during which "myriads of different ... time cultures" (Osterhammel 2009: 119 [my translation])[20] were reduced to a diminished diversity. The German historian Jürgen Osterhammel argues that the plurality of time cultures was reduced by the late nineteenth century because of several successful initiatives to coordinate techniques of time measurement on both an internal as well as an international level in most of the Western industrialised countries (ibid., 119).[21] In fact, in 1883 the United States defined and introduced a national standard time with four time zones that was based on the British Greenwich meridian (Geyer 2005: 95; McCrossen 2007: 222). One year later, the International Meridian Conference took place in Washington, DC, where 25 delegates from different countries discussed and ultimately approved of the adoption of a system of time measurement that was valid on a global scale. Greenwich became the universally accepted prime meridian (McCrossen 2007: 223).[22] In the German

---

[18]"Menschen des Übergangs"; "die Gegenwart in einem großen Prozess der Umwandelung [sic]".

[19]This coincides with assumptions one finds in the introduction to Stephen Kern's ground-breaking study *The Culture of Time and Space 1880–1918* (1983: 6). Still, the question of what has actually been interpreted and defined as being 'the present' in different historical contexts—in this case the German *Kaiserreich* around 1900—has hardly been examined up to now. This is a desideratum that should trigger further historiographical research, especially if one takes into consideration that the present is the only dimension of time in which people actually dispose of it as a resource. It is exclusively in the present that people can decide how they are going to evaluate and organise time, cf. Landwehr (2014: 37, 179).

[20]"Myriaden unterschiedlicher ... Zeitkulturen".

[21]It was especially the railroad and its timetables that made the need for time coordination obvious.

[22]The globe was divided up into a total of 24 time zones, each one of them encompassing 15 degrees of longitude. In France, however, for example, the adoption of the universal time standard was delayed until 1911, mostly the result of issues involving national pride (Geyer 2005: 96; Osterhammel 2009: 120–121).

Kaiserreich, the military had argued strongly in favour of introducing a national standard time. Here, a corresponding law was passed in 1893 (Dohrn-van Rossum 1995: 320; Geyer 2005: 96; Landwehr 2014: 30; Osterhammel 2009: 120–121).[23]

Even before these synchronised references to and usages of time were established on both a national and a global level, a process that Osterhammel labelled "chronometrisation" (Osterhammel 2009: 121 [my translation])[24] had already become evident in the Western industrialised countries since the mid-nineteenth century. On the one hand, this term takes into account the increasingly vast dissemination of timekeepers for private use because of the possibilities of industrial mass production (ibid., 122). According to Osterhammel, privately owned clocks became "one of the most potent symbols of modernity" (ibid., 125 [my translation]).[25] Their dissemination reinforced the heightened awareness for time in general and involved punctuality becoming a virtue (ibid., 122). It was to this interrelationship that the German historian Karl Lamprecht referred in 1912:

> What initially characterises the modern concept of time is the precise and hands-on consideration of the small period of time: five-minute audiences, minute-talks on the telephone, the rotary printing press' production by the second, fifth-of-a-second measurements for bike riding—put morally: punctuality. ... At the same time, the perception of what time is and what can be measured with regard to time has actually grown. ... (Lamprecht quoted in Flemming et al. 1997: 42–43 [my translation])[26]

In the same context, Lamprecht explained the widespread demand for pocket watch accuracy down to the second with people's longing for a better overview and more agency insofar as daily life routines were

---

[23] In the German Reichstag in 1891, Helmuth von Moltke had opposed the plurality of local times to the necessity of rapidly mobilising the army in case of emergency (Geyer 2005: 97–98).

[24] "Chronometrisierung".

[25] "zu einem der potentesten Symbole von Modernität".

[26] "Was also den modernen Zeitbegriff zunächst kennzeichnet, das ist die genaue praktische Beachtung des kleinen Zeitabschnittes: Fünfminutenaudienzen, Minutengespräche am Telephon, Sekundenproduktion der Rotationsdruckmaschine, Fünftelsekundenmessung beim Fahrrad: moralisch ausgedrückt Pünktlichkeit. ... Zugleich aber ist damit der Sinn für das, was Zeit ist und sich an Zeit messen läßt, überhaupt gewachsen ... ."

concerned (ibid.), an explanation that conversely points to the presence of unknown time experiences.

On the other hand, public clocks continued to be constructed and regarded as national symbols, as their installation was meant not only to make time visible and audible but also to fulfil purposes of political representation (Osterhammel 2009: 123; McCrossen 2007: 217). According to the historian Alexis McCrossen, "by the end of the nineteenth century, public clocks in the United States reached the peak of their practical and symbolic importance" (McCrossen 2013: 7). Government-sponsored public clocks were supposed to fulfil purposes of nationalist symbolism in the United States, especially after the end of the Civil War (McCrossen 2007: 217, 225). In Prussia, a number of official decrees issued in the course of the nineteenth century are evidence of the administration's growing efforts to bring the diversity of local times under some sort of standardised control (Geyer 2005: 87): "The clocks of the local post offices were established as authoritative local 'master clocks,' thus dethroning the clocks in church steeples and city towers ..." (ibid., 88). In Berlin, the first so-called *Normalsekundenuhr*, a public master clock whose novelty significantly lay in its capacity to "be precise down to a second" (ibid., 92), was erected in 1869. Public clocks thus equally met the widespread wish for greater temporal coordination and punctuality, which proved indispensable, especially for modern business life in the city. As the German philosopher and sociologist Georg Simmel wrote in his famous essay "The Metropolis and Mental Life" (*Die Großstädte und das Geistesleben*), first published in 1903:

> The relationships and concerns of the typical metropolitan resident are so manifold and complex that, especially as a result of the agglomeration of so many persons with such differentiated interests, their relationships and activities intertwine with one another into a many-membered organism. In view of this fact, the lack of the most exact punctuality in promises and performances would cause the whole to breakdown into an inextricable chaos. If all the watches in Berlin suddenly went wrong in different ways even only as much as an hour, its entire economic and commercial life would be derailed for some time. ... For this reason the technique of metropolitan life in general is not conceivable without all of its activities and reciprocal relationships being organized and coordinated in the most punctual way into a firmly fixed framework of time which transcends all subjective elements. (Simmel [1903] 2010: 105)

However, precisely these public clocks were often out of synchrony. The technically successful implementation of regulating systems, which guaranteed the synchronicity of public clocks, became a fact only at the beginning of the twentieth century. Thus, complaints about the unreliability of public clocks were a recurrent phenomenon in many cities in the United States and in Europe around 1900 (Geyer 2005: 98; McCrossen 2013: 12–13, 114–115, 144–145). In Berlin, bicyclists made it a sport to check whether the public clocks were running accurately. In fact, the director of the Royal Observatory officially invited the Berliners to report any inaccuracy (Geyer 2005: 93), an invitation which sensed that there still might be pluritemporality. What is particularly striking in Simmel's statement in this context, then, is the accent he puts on the plurality of daily modern city life. It is this diversity that triggered the assumed need for temporal synchronisation in the first place. Thus, the plead for simultaneity has to be regarded as a *reaction* to the manifold, confusing, and unfamiliar uses of present time(s) in the era of "high modernity." Furthermore, the undoubtedly increasing endeavours to ensure temporal synchronisation initially did nothing but multiply the possible combinations of this 'contemporaneity of the noncontemporaneous' (Ogle 2015). Different experiences and uses of time in everyday life obviously collided with a growing number of attempts to coordinate and regulate time on an institutional level (ibid., 99). The latter, driven by political or economic intentions, prominently highlighted power-related ways of handling time; this again led to more and more discussions of time-related issues, either consciously or unconsciously so.[27] Synchronised simultaneity was not a fact that completely dominated daily life at the turn of the nineteenth to the twentieth century in the German *Kaiserreich*. Rather, present time, its conceptions and usages, appeared to diversify constantly to the point where it seemed difficult, if not impossible, to become familiar with all its facets. The present, therefore, was unknown and a highly contested dimension in Germany around 1900.

---

[27] This connects to Daniel E. Agbiboa's claim that comments about unknown time serve as a way of dealing with the uncertainties and insecurities experienced in daily life; see Sect. 'No-Place-Land' and 'Neverland'.

PRESENT TIME(S) AND AMERICANISATION

Society around 1900 was not only influenced by underlying temporal structures, but people actively participated in the construction of temporal discourses. In this context, Americanisation had a crucial role. The term *Americanisation* was used to refer to the intellectual, technological, or cultural transfer processes from the United States to the 'old world,' regardless of whether these transfer processes actually took place or were just implied.[28] For many Germans at the time, Americanisation was a convenient rhetorical instrument that served to highlight and comment on certain processes of modernisation. Cultural pessimists, such as the art critic Karl Scheffler, often referred to the asynchronicity they observed in cities such as Berlin as just one example of the overall decline of traditional, long-established 'German culture' in the context of modernity. To Scheffler, modernisation was Americanisation and led to the implementation of values linked to 'international civilisation' (Scheffler 1910: 178).[29] Obviously, these were values of which Scheffler disapproved. From his perspective, American-based civilisation meant worshipping quantity, and perhaps technical superiority, as the highest value(s) of all, but excluded historical continuities and brought about cultural decay. However, Scheffler's expressions of opinion were ambivalent: He did recognise the economic efficiency that unfolded in contemporary Berlin and evaluated it positively, although this was a development whose characteristics he believed to originate in North America (Scheffler 1910: 171). In this sense, Scheffler's statements reflect not only an aversion towards modern pluritemporality and its unknown components but equally show a common fascination for certain uses of time with which Germans were unfamiliar and that they attributed to 'the Americans.' In fact, in 1904, the philosopher and psychologist Hugo

---

[28] "The British journalist W.T. Stead coined the phrase 'Americanization of the world' in his 1902 book that in alarmist tones detailed the impact of America on Europe and the British Empire" (Nolan 2012: 30).

[29] The juxtaposition of 'culture' on the one hand and 'civilisation' on the other is a typical one for intellectual and above all cultural pessimistic comments of the time. Later on, it also served as a rhetorical legitimation for the supposed necessity of fighting for 'German culture' and against 'Western civilisation' in World War I (Bollenbeck 1999: 18–27; Kroll 2003: 9).

Münsterberg published a study in two volumes entitled *Die Amerikaner* (The Americans), in which he offered extensive descriptions of what he had experienced to be 'typically American'.[30] Significantly, it is, amongst other things, the way Americans organised their time that fascinated him:

> This manifests itself especially in the downright virtuosic utilisation of time. Superficial observers have sometimes misinterpreted, suggesting that Americans were always in a hurry, but just the opposite is true. Somebody who needs to rush all the time has made bad use of his time and therefore does not have the full time necessary for the accomplishment of his work in a perfect manner. The American is not in a hurry at all. He rather makes use of his time in a way that doesn't generate any waste of this valuable material at all. He does not like to wait or to be idle. One task accurately follows the next, and one thing after another is being executed smoothly with factual precision. (ibid., 374–375 [my translation])[31]

What obviously gained Münsterberg's admiration is an absolutely rational use of time, which he perceived to be a 'typically American' trait. He emphasised that efficient time management has to be distinguished from haste and hurry. German businessman Ludwig Max Goldberger shared these evaluations when he recalled his experiences in the United States in his book *Das Land der unbegrenzten Möglichkeiten* (The Land of Endless Possibilities)—a title that has coined a slogan since 1903:

> Speaking of a nervousness of commercial haste on the other side of the ocean is wrong. The opposite is true. It is simply endless activity one

---

[30] At the time he wrote this study, Münsterberg had already lived and worked in the United States for about ten years, amongst others at Harvard University in Boston (Münsterberg [1904] 1912: Preface).

[31] "Ganz besonders bekundet sich das in der geradezu virtuosenhaften Ausnützung der Zeit. Oberflächliche Betrachter haben das manchmal so mißdeutet, als wenn der Amerikaner stets in Eile sei, aber gerade das Gegenteil ist der Fall. Wer fortwährend eilen muß, hat über seine Zeit schlecht disponiert und hat somit für nichts die volle Zeit, die nötig ist, um die Arbeit in vollendeter Weise zu erledigen. Der Amerikaner ist durchaus nicht in Eile. Dagegen trifft er die Verfügung über seine Zeit so, daß nirgends Vergeudung dieses kostbaren Materials entsteht. Er mag nicht warten oder müßig sein. Eines gliedert sich genau ans andere an und mit sachlicher Präzision wird eins nach dem anderen glatt erledigt."

perceives there, intense hard work, and over and over again hard work. ... (Goldberger [1903] 1911: 29 [my translation])[32]

Picking up the concept of 'nervousness' in this context is by no means coincidental: the complaints of contemporaries about speeding and hasting led to a new mass phenomenon, if not a mass epidemic, namely that of nervousness. These complaints first came up in the United States and shortly afterwards in the German *Kaiserreich*. Nervousness was not only a real suffering but also a cultural construct and as such just one of the many temporal discourses that, in addition to acceleration, were contemporarily up to date (Radkau 1998: 10–13; Eckart 2009).[33]

The Münsterberg and Goldberger quotes also show in what way America's present was seen as Germany's possible and desirable future (Lüdtke et al. 1996: 7, 10): Their praises of the American willingness to work hard and efficiently went hand in hand with an image of the United States as a nation that was 'man-made'. Indeed, many Germans saw Americans positively as 'men of action' having constructed a society where 'time is money' (Nolan 2012: 34).[34] Especially after 1903, when the American mechanical engineer Frederick Winslow Taylor had published his first book *Shop Management*, German contemporaries considered 'typically American' ways of dealing with time as rooted in Taylorism. This ostensibly scientific programme of organising and regulating work routines according to the criteria of time was linked to rationality and efficiency—qualities that were understood as ultimately leading to general prosperity (Maier 1970: 29; Nolan 1994: 42).

Some intellectuals further perceived the United States of America as a nation that was not—in contrast to the German-speaking nations— lost in asynchronicity. Austrian architect Adolf Loos, for example, wrote in his essay "Ornament und Verbrechen" (Ornament and Crime) from 1908:

---

[32] "Man spricht mit Unrecht von einer Nervosität des erwerblichen Hastens auf der andern [*sic*] Seite des Ozeans. Das Gegenteil ist der Fall. Nur unendliche Regsamkeit nimmt man dort wahr, angestrengten Fleiß und immer wieder Fleiß ... ."

[33] Another example of a temporal discourse of the time would be the cultural pessimistic notion of a *fin de siècle* that was widespread in the educated middle class by the end of the nineteenth century.

[34] The statement that 'time is money' can be traced back to Benjamin Franklin and the year 1748 (Schwarz 2001: 465).

The speed of cultural development suffers from latecomers. I might live in the year 1908, but my neighbour lives around 1900 and another one in the year 1880. It is a disaster for a nation if the culture of its inhabitants is spread over such a big period of time. ... A country can call itself lucky if it doesn't have such latecomers and marauders. Lucky America! (Loos [1908] 2016 [my translation])[35]

According to Loos, America could proudly claim to be the land of simultaneity, whereas German-speaking nations of this period were considered to be pitiable sites of asynchronicity, not only by him but also by many of his contemporaries (Geyer 2007: 177). To overcome the plurality of both public and private times in modern daily life, for example by finding possibilities to create the sensation of shared experiences, was thus thought to be a primary necessity to ensure a stable political and social community.

In the German *Kaiserreich*, even the official answer to the question of when the century was actually meant to turn can be interpreted in that way: as an attempt to create a nationally shared experience of this "big turning point in time". Most German intellectuals considered the twentieth century to start on January 1, 1901, but the popular conception was to localise the beginning of the new century exactly one year earlier. According to his intention to strengthen the image of unified nationhood, Kaiser Wilhelm II. opted for the popular conception. Thus, the Reichstag issued an official decree about the start of the twentieth century on January 1, 1900.[36] In the light of the argument that has been made in this essay, this decision proves that perceptions of 'the present' were obviously so diverse—and thus unknown—in Germany around 1900 that the highest state institutions felt the need to exert their power and to define precise points in time.

---

[35] "Das tempo der kulturellen entwicklung leidet unter den nachzüglern. Ich lebe vielleicht im jahre 1908, mein nachbar aber lebt um 1900 und der dort im jahre 1880. Es ist ein unglück für einen staat, wenn sich die kultur seiner einwohner auf einen so großen zeitraum verteilt. ... Glücklich das land, das solche nachzügler und marodeure nicht hat. Glückliches Amerika!"

[36] Still, most of the other European nations as well as the United States opted for the correct point of time as measured by scientific standards and celebrated the beginning of the new century one year later than the Germans (Brendecke 1999: 224–225; Frevert 2000: 8–10; Puschner 2010: 528–529; Salewski 1971: 338–341).

Bibliography

Adams, Henry. [1907] 2011. *The Education of Henry Adams.* Project Gutenberg. http://www.gutenberg.org/files/2044/2044-h/2044-h.htm.

Bollenbeck, Georg. 1999. *Tradition, Avantgarde, Reaktion: Deutsche Kontroversen um die kulturelle Moderne, 1880–1945.* Frankfurt am Main: S. Fischer.

Brendecke, Arndt. 1999. *Die Jahrhundertwenden: Eine Geschichte ihrer Wahrnehmung und Wirkung.* Frankfurt am Main: Campus.

Dipper, Christof. 2000. Die 'Geschichtlichen Grundbegriffe': Von der Begriffsgeschichte zur Theorie der historischen Zeiten. *Historische Zeitschrift* 270: 281–308.

———. 2010. Moderne. *Docupedia-Zeitgeschichte.* http://docupedia.de/zg/Moderne. Accessed 10 Feb 2016.

Dohrn-van Rossum, Gerhard. 1995. *Die Geschichte der Stunde: Uhren und moderne Zeitordnungen.* München: Carl Hanser.

Eckart, Wolfgang U. 2009. Nervös in den Untergang: Zu einem medizinisch-kulturellen Diskurs um 1900. *Zeitschrift für Ideengeschichte* 3 (1): 64–79.

Flemming, Jens, Klaus Saul, and Peter-Christian Witt (eds.). 1997. *Quellen zur Alltagsgeschichte der Deutschen: 1871–1914.* Darmstadt: Wissenschaftliche Buchgesellschaft.

Frevert, Ute. 2000. Jahrhundertwenden und ihre Versuchungen. In *Das Neue Jahrhundert: Europäische Zeitdiagnosen und Zukunftsentwürfe um 1900,* ed. Ute Frevert, 7–14. Göttingen: Vandenhoeck & Ruprecht.

Geppert, Alexander C. T., and Till Kössler. 2015. Zeit-Geschichte als Aufgabe. In *Obsession der Gegenwart: Zeit im 20. Jahrhundert,* eds. Alexander C. T. Geppert and Till Kössler, 7–36. Göttingen: Vandenhoeck & Ruprecht.

Geyer, Martin H. 2005. Prime Meridians, National Time, and the Symbolic Authority of Capitals in the Nineteenth Century. In *Berlin-Washington, 1800–2000: Capital Cities, Cultural Representation, and National Identities,* eds. Andreas W. Daum and Christof Mauch, 79–100. Cambridge: Cambridge University Press.

———. 2007. 'Die Gleichzeitigkeit des Ungleichzeitigen': Zeitsemantik und die Suche nach Gegenwart in der Weimarer Republik. In *Ordnungen in der Krise: Zur politischen Kulturgeschichte Deutschlands, 1900–1933,* ed. Wolfgang Hardtwig, 165–187. München: Oldenbourg.

Goldberger, Ludwig Max. [1903] 1911. *Das Land der unbegrenzten Möglichkeiten: Beobachtungen über das Wirtschaftsleben der Vereinigten Staaten von Amerika.* Berlin: Fontane.

Graf, Rüdiger. 2012. Zeit und Zeitkonzeptionen in der Zeitgeschichte. *Docupedia-Zeitgeschichte.* http://docupedia.de/zg/Zeit_und_Zeitkonzeptionen_Version_2.0_R%C3%BCdiger_Graf. Accessed 10 Feb 2016.

Gumbrecht, Hans Ulrich. 1978. Modern: Modernität, Moderne. In *Geschichtliche Grundbegriffe: Historisches Lexikon zur politisch-sozialen Sprache in Deutschland*, vol. 4, eds. Otto Brunner, Werner Conze, and Reinhart Koselleck, 93–131. Stuttgart: Klett-Cotta.

Herbert, Ulrich. 2007. Europe in High Modernity: Reflections on a Theory of the 20th Century. *Journal of Modern European History* 5: 5–20.

Hübinger, Gangolf. 2014. Wissenschaften, Zeitdiagnosen und politisches Ordnungsdenken. Zur Einführung. In *Europäische Wissenschaftskulturen und politische Ordnungen in der Moderne: 1890–1970*, ed. Gangolf Hübinger, 1–28. München: Oldenbourg.

Jung, Theo. 2010–2011. Das Neue der Neuzeit ist ihre Zeit: Reinhart Kosellecks Theorie der Verzeitlichung und ihre Kritiker. *Moderne: Kulturwissenschaftliches Jahrbuch* 6: 172–184.

Kern, Stephen. 1983. *The Culture of Time and Space: 1880–1918*. Cambridge: Harvard University Press.

Koselleck, Reinhart. 1972. Introduction. *Geschichtliche Grundbegriffe: Historisches Lexikon zur politisch-sozialen Sprache in Deutschland*, vol. 1, eds. Otto Brunner, Werner Conze, and Reinhart Koselleck, xiii–xxvii. Stuttgart: Klett-Cotta.

———. 1979. *Vergangene Zukunft: Zur Semantik geschichtlicher Zeiten*. Frankfurt am Main: Suhrkamp.

———. 2000. Gibt es eine Beschleunigung der Geschichte? In *Zeitschichten: Studien zur Historik*, 150–177. Frankfurt am Main: Suhrkamp.

———. 2004. *Futures Past: On the Semantics of Historical Time*. New York: Columbia University Press.

Kroll, Frank-Lothar. 2003. *Kultur, Bildung und Wissenschaft im 20. Jahrhundert*. München: Oldenbourg.

———. 2013. *Geburt der Moderne: Politik, Gesellschaft und Kultur vor dem Ersten Weltkrieg*. Bonn: Bundeszentrale für politische Bildung.

Landwehr, Achim. 2014. *Geburt der Gegenwart: Eine Geschichte der Zeit im 17. Jahrhundert*. Frankfurt am Main: Fischer.

Loos, Adolf. [1908] 2016. Ornament und Verbrechen (Auszüge). In *Architekturtheorie*, ed. Technische Universität Berlin/Institut für Architektur. *Technische Universität Berlin*. https://www.architekturtheorie.tu-berlin.de/fileadmin/fg274/1_-_Adolf_Loos__Ornament_und_Verbrechen__1908_-_Auszug.pdf. Accessed 9 Feb 2016.

Lüdtke, Alf, Inge Marßolek, and Adelheid von Saldern. 1996. Amerikanisierung: Traum und Alptraum im Deutschland des 20. Jahrhunderts. In *Amerikanisierung: Traum und Alptraum im Deutschland des 20. Jahrhunderts*, eds. Alf Lüdtke, Inge Marßolek, and Adelheid von Saldern, 7–33. Stuttgart: Steiner.

Maier, Charles S. 1970. Between Taylorism and Technocracy: European Ideologies and the Vision of Industrial Productivity in the 1920s. *Journal of Contemporary History* 5: 27–51.

McCrossen, Alexis. 2007. Conventions of Simultaneity: Time Standards, Public Clocks, and Nationalism in American Cities and Towns, 1871–1905. *Journal of Urban History* 33 (2): 217–253.

———. 2013. *Marking Modern Times: A History of Clocks, Watches, and Other Timekeepers in American Life*. Chicago: The University of Chicago Press.

Middendorf, Stefanie. 2012. Mass Culture as Modernity: Introductory Thoughts. *Journal of Modern European History* 2: 147–151.

Münsterberg, Hugo. [1904] 1912. *Die Amerikaner*, vol. 1: Das politische und wirtschaftliche Leben. Berlin: Mittler.

Nolan, Mary. 1994. *Visions of Modernity: American Business and the Modernization of Germany*. New York: Oxford University Press.

———. 2012. *The Transatlantic Century: Europe and America, 1890–2010*. Cambridge: Cambridge University Press.

Ogle, Vanessa. 2013. Whose Time Is It? The Pluralization of Time and the Global Condition, 1870s–1940s. *American Historical Review* 118 (5): 1376–1402.

———. 2015. *The Global Transformation of Time: 1870–1950*. Cambridge: Harvard University Press.

Osterhammel, Jürgen. 2009. *Die Verwandlung der Welt: Eine Geschichte des 19. Jahrhunderts*. München: C.H. Beck.

Peukert, Detlev J.K. 1987. *Die Weimarer Republik: Krisenjahre der klassischen Moderne*. Frankfurt am Main: Suhrkamp.

Puschner, Uwe. 2010. Rückblicke, Vorblicke: Krisenbewußtsein und Umbruchserfahrung im Augenblick der Jahrhundertwende. In *Krisenwahrnehmungen in Deutschland um 1900: Zeitschriften als Foren der Umbruchszeit im Wilhelminischen Reich*, eds. Michel Grunewald and Uwe Puschner, 525–536. Bern: Peter Lang.

Radkau, Joachim. 1998. *Das Zeitalter der Nervosität: Deutschland zwischen Bismarck und Hitler*. München: Carl Hanser.

Raphael, Lutz. 2008. Ordnungsmuster der 'Hochmoderne'? Die Theorie der Moderne und die Geschichte der europäischen Gesellschaften im 20. Jahrhundert. In *Dimensionen der Moderne: Festschrift für Christof Dipper*, eds. Ute Schneider and Lutz Raphael, 73–91. Frankfurt am Main: Peter Lang.

Rosa, Hartmut. 2005. *Beschleunigung: Die Veränderungen der Zeitstrukturen in der Moderne*. Frankfurt am Main: Suhrkamp.

———. 2013. *Beschleunigung und Entfremdung: Entwurf einer Kritischen Theorie spätmoderner Zeitlichkeit*. Bonn: Bundeszentrale für politische Bildung.

Salewski, Michael. 1971. 'Neujahr 1900': Die Säkularwende in zeitgenössischer Sicht. *Archiv für Kulturgeschichte* 53: 335–381.

Scheffler, Karl. 1910. *Berlin: Ein Stadtschicksal.* Berlin: Reiss.

Schwarz, Angela. 2001. 'Wie uns die Stunde schlägt': Zeitbewußtsein und Zeiterfahrung im Industriezeitalter als Gegenstand der Mentalitätsgeschichte. *Archiv für Kulturgeschichte* 83 (2): 451–479.

Seefried, Elke. 2015. *Zukünfte: Aufstieg und Krise der Zukunftsforschung, 1945–1980.* Berlin: De Gruyter.

Simmel, Georg. [1903] 2010. The Metropolis and Mental Life, trans. Edward Shils. In *The Blackwell City Reader*, eds. Gary Bridge and Sophie Watson, 2nd ed, 103–110. Chichester: Blackwell.

Stockhorst, Stefanie. 2006. Zeitkonzepte: Zur Pluralisierung des Zeitdiskurses im langen 18. Jahrhundert. *Das achtzehnte Jahrhundert* 30 (2): 157–164.

PART III

Future Pasts

CHAPTER 11

The Time-Image and the Unknown in Wong Kar-wai's Film Art

Dorothee Xiaolong Hou and Sheldon H. Lu

Hailed as a leading art-house director in world cinema, Hong Kong director Wong Kar-wai (王家卫) is unique in the representation of filmic time. His films are characterised by episodic narration, unique cinematography, multiple speeds and temporalities (normal tempo, slow motion, and accelerated tempo), blurry images, and saturated colours. These are not conventional narratives of the "movement-image" but excel in the depiction of the "time-image" (Deleuze 1989: 98), in which time does not simply help develop plot and movement but is itself at the centre. Wong Kar-wai's films display a strong fascination with time and temporality, their construction, deconstruction, fragmentation, reconstruction, projection, and retrospection. Memory, nostalgia, and amnesia are recurrent motifs in Wong's films. Wong explores the artificiality and complexity of memories, the alienation of historical time, as well as anxieties of the merging of individual and collective memories. As we argue here, Wong's films embody the principles of the "time-image" in modern

D.X. Hou (✉) · S.H. Lu
University of California, Davis, CA, USA

© The Author(s) 2017
S. Baumbach et al. (eds.), *The Fascination with Unknown Time*,
DOI 10.1007/978-3-319-66438-5_11

cinema, or more appropriately, postmodern cinema, which is a cinema of fragmentation, self-reflexivity, paradox, double and multiple temporalities, nonlinear narrative, and inconclusive endings. Using Gilles Deleuze's concepts of "movement-image" and "time-image," we explore various configurations of time in Wong Kai-wai's films. Thereby, we focus especially on Wong's informal trilogy, namely, *Days of Being Wild* (1990), *In the Mood for Love* (2000), and *2046* (2004), in which Wong uses exemplary time-images to explore problems of time and memory in the changing geopolitical and psychosocial landscapes of Hong Kong, attempting to open up a space for its past to negotiate with its unknown future.

THE MOVEMENT-IMAGE AND THE TIME-IMAGE

As claimed by Deleuze, before Henri Bergson, time was seen as the sequence of similar events that take place in a series of moments set in a linear fashion. In this context, time is represented, for example, as a clock. For Deleuze, this traditional conception of time does not reveal time's true and entire nature but only offers the spatialisation of time, which is problematic. Modern conceptions of time, on the other hand, conceive of time in its pure form, as lived by man. Instead of depicting time as moving forward horizontally (on an 'x-axis'), time was imagined as vertical (on a 'y-axis'), because man can expand time by layering the present with the past and the future. As stated by Deleuze, activities of recognising, recollecting, and dreaming further expand this 'y-axis.' Deleuze's understanding of time is a phenomenological one: the past and the future (the 'virtual') coexist with the present (the 'actual'), wherein the former penetrates into the latter in the form of memory or desire. This notion of time is the basis for the concept of time-image in cinema.

Another element crucial to Deleuze's philosophy of film is the concept of *image*. A Deleuzian view of the world assumes that the way we live our lives is very similar to filmmaking. For each perception, action, and affection in which we engage, we frame and image a relationship of sorts: a relationship between the self as an entity and the universe. For Deleuze, *image* is less of a noun than a verb: it refers to the process of capturing, reflecting, and refracting the world. Therefore, a Deleuzian reading of films is not so much concerned with the mimetic function of cinematic images, that is, with cinema as a somewhat representational medium. Instead, Deleuze reads cinema as *it is*, demonstrating various ways of "worldmaking" in which it engages (Goodman 1978: 7).

With regard to the 'image,' Deleuze distinguishes between two different categories, the movement-image and the time-image. The movement-image—or more precisely, 'imaging the movements' (*l'image-mouvement*)—is not an image that moves, but an image of "pure mobility extracted from the movements of characters" (Deleuze 1986: 25). Connecting to Bergson's critique on 'clock time,' the movement-image, according to Deleuze, provides an "indirect image of time" (Deleuze 1989: 22). He groups the movement-image into three main categories: the perception-image, the affection-image, and the action-image, which operates as a sensory motor system, where action and reaction, cause and result are linked through the progression of time and thus form a broadly linear plot in films (Powell 2007: 145). The time-image (*l'image-temps*), on the other hand, provides a direct image of time. It is no longer subordinate to movement but presents "time in [its] pure state" (Deleuze 1989: 169). In contrast to the movement-image, the time-image embodies an openness of time, wherein the past and future are not linked in a linear fashion: instead, they coexist and intermingle in a state of constant renewal. In films of the time-image, the virtual and the actual merge in an imperceptible manner. In Wong Kar-wai's films, the time-image is created by matching and mirroring images and characters. Film narrative tends to oscillate indiscernibly between the past or future and the present.

THE CRISIS OF THE MOVEMENT-IMAGE: WONG'S URBAN LOVE TALES

The directorial debut of Wong Kar-wai (born in Shanghai, 1958), *As Tears Go By* (旺角卡门, 1988), is a crime melodrama about three young people: Wah (Andy Lau), a mobster; his friend and apprentice Fly (Jacky Cheung), who is ambitious but often gets into trouble; and Ngor (Maggie Cheung), Wah's younger cousin, who later becomes his love interest. As much as Wah wants to have a 'normal' and peaceful life with Ngor, he is constantly involved in mob violence to protect Fly. In the end, Wah is shot on an assassination mission and, as he is gunned down, he has a flashback of his first kiss with Ngor.

Often overlooked, this film lays the foundation of Wong's impending emergence as an art-house auteur. Taking on the popular theme of Hong Kong triads, Wong's film, however, does not follow the common theatrical bent of its kind. In this film, the often glorified fraternal bond of duty—a clichéd theme in Hong Kong gangster films—is downplayed;

the protagonists do not have any clear motivation or purpose; and the ending of the film—Wah's death—only vaguely resembles a 'closure' (Rushton 2012: 145), as it does not offer a conclusion nor a moral lesson for the audience. This unconventional narrative pattern, with only a minimal sense of progression, and the fascination with the quotidian life of gangsters, can be seen as a divergence from the Deleuzian "large form of the action-image" (Deleuze 1986: 155). The latter is defined, in Deleuze's *Cinema 1*, by the narrative movement of what he calls "situation-action-situation," or SAS' form. In films of "the large form," the initial situation (S) calls for, and is in turn changed by certain acts (A), which leads to the emergence of the latter situation (S'). This "large form" requires "a gap between the situation and the action to come, but this gap only exists to be filled" (ibid., 155). In *As Tears Go By*, Wah finds it increasingly difficult to maintain his relationship with Ngor while trying to protect Fly. We see Wah's growing dilemma of being either too late or too early, never being able to act between situations and resolve problems. If we use Deleuze's term, what we see is the closing gap between situations (S and S') and the decreasing mobility (A) of the characters. Hence we see the crisis of the "large form:" the impossibility of love for a small-time gangster such as Wah. It is less a result of personal failure than a formal one.

In his later films, *Chungking Express* (重庆森林, 1994) and *Fallen Angels* (堕落天使, 1995), often considered as two companion pieces, Wong continues to explore the theme of failed love in the big city. *Chungking Express* is about the excitement of switching roles and exploring the unknown future of its youthful protagonists. It is a film about the projection of future possibilities, marked by nonlinear narration, fragmentation, and open-endedness. It is a film about the stories of two Hong Kong police officers. The first part of the film is the story of Cop 223, He Qiwu, nicknamed Ah Wu (Takeshi Kaneshiro in Japanese, Jin Chengwu in Chinese) and his platonic encounter with a mysterious drug smuggler (Brigitte Lin). The second part of the film is the story of Cop 663 (Tony Leung Chiu-wai) and his relationship with his erstwhile flight attendant-girlfriend, and store owner-turned-flight attendant Faye (Faye Wong). The film portrays role-playing among young people and their fantasising about different lifestyles and future possibilities. *Fallen Angels*, darker and more sinister than *Chungking Express*, is about the encounters and adventures of five young people in Hong Kong's nightlife. The film also consists of two separate but loosely connected stories. The first story is about an uncanny love triangle between a hit man, Wong Chi-Ming (Leon Lai), his 'partner in crime' (Michelle Reis), and a wild prostitute, Blondie (Karen Mok).

In *As Tears Go By*, we see the difficulty of the "big form of action-image" in portraying the capriciousness of life in the hustle and bustle of the metropolis. In comparison, in *Chungking Express* and *Fallen Angels*, we see a "small form of action-image" come into play. According to Deleuze, although the large form follows the SAS' form, the small form follows the ASA' form. The former is based on moving from one situation to an action, which modifies the prior situation into a new one; the latter involves moving from an action to a situation, then to a new action. With regard to the ASA' form, Deleuze writes: "the action advances blindly and the situation is disclosed in darkness, or in ambiguity. From action to action, the situation gradually emerges, varies, and finally either becomes clear or retains it mystery. ... It is no longer ethical, but comedic" (Deleuze 1986: 160). For example, in *Fallen Angels*, Ah Wu, the light-hearted delinquent, makes a living—and also enjoys it as a hobby—by breaking into small businesses at night and pretending to be the owner, often offering people goods and services by force. He meets a neurotic girl, Charlie (Charlie Yeung), and goes on mischievous adventures with her. The film shows the two characters stumble from one action to another, without any coherence or any tangible goal. In the action-image, the ethical imperative—the struggle between the good and the evil—loses its gravity and is replaced by a series of endless, often meaningless actions. Displayed as a series of temporal ellipses, in slow motion, and in freeze-frame, memory is depicted as malleable while the future is unpredictable.

The fragmentation of narrative-time illustrates the urban experience, the catastrophe of love in the metropolis—a theme frequently visited by Wong Kar-wai. In lieu of creating a conventional love story, the film portrays individuals strolling in the city and encountering one another by pure chance, evoking a Baudelairean 'stigmata' that the big city inflicts upon love. Benjamin refers to it as "love at last sight," a farewell that forever coincides with the very moment of enchantment (Benjamin 1968: 169).

From Movement-Image to Time-Image: The *2046* Trilogy

Days of Being Wild, In the Mood for Love, and *2046* are sometimes regarded as Wong Kai-war's informal trilogy. The first movie in this celebrated trilogy, *Days of Being Wild* (阿飞正传, 1990), is about the dissolute life of a playboy called Yuddy (Leslie Cheung) in Hong Kong in 1960. First, he courts Su Li-zhen (Maggie Cheung), a reserved shop clerk, but ends up leaving her when she asks for his commitment. He then pursues Mimi (Carina Lau), a shrewish but vivacious cabaret dancer, but dumps her, too. Adopted by a former courtesan (Rebecca Pan), Yuddy never knew his biological mother and desperately wants to meet her. Eventually, he leaves Hong Kong for the Philippines to look for her but gets rejected. Disappointed and disillusioned, Yuddy clashes with local gangsters and is shot dead on a train. In this sentimental and melancholy love elegy, Wong Kar-wai questions and ultimately denies the possibility of retaining the past by virtue of recollection. The film thus serves as a critique on the pre-Bergsonian clock-time.

At the beginning of the film, Yuddy approaches Su, asking her to look at his watch while sharing one minute of companionship with him. Then he tells her: "From now on, we will only be 'friends for one minute.' You cannot deny it, because it (this minute) has already passed" (*Days of Being Wild*, directed by Wong Kar-wai, 1990). Later in the film, he says the same thing to Mimi, when he refuses to be committed to their relationship. These conversations (or, rather, Yuddy's monologues), cutting back and forth to the clock-image, appear repeatedly in a nearly compulsive manner throughout the film. Just like Su and Mimi, the audiences are seduced by Yuddy's tantalising monologues, which appear to evade and even defy clock-time, even though in the end none of the characters in the film can escape it. In this film, we see again the "small form of action-image," that is, the ASA' form: Yuddy would pursue a woman with similar modes of actions—telling her that they are "friends for one minute" or the story of "a feetless bird" (ibid.)—and then leave her and move on to another woman. Here, the equivocal action (A)—Yuddy's charming act of seduction, if we may call it that—culminates in a clarified situation (S), where women fall for him and become his lovers. This situation then makes possible a new action (A'), where Yuddy, a "feetless bird" indeed, roams towards his next target. Yuddy's repetitive seduction is what Deleuze calls an 'index of lack' (*indice de manque*). It implies "a hole in the story," a missing situation that is only gradually revealed

(Bogue 2003: 89). As the story unfolds, we know it is the trauma of being deserted by his biological mother that causes Yuddy's dilemma between his longing for intimacy and his inability to love. What is also revealed is the crisis of the movement-image: The image of the clock, in which Yuddy's blandishments are compressed into 'one minute,' is not really an image of movement: it is an image of thought, of Yuddy's endless and recurring trauma. The tragedy of the feetless bird points to the plight of clock-time: time cannot be fixed to definiteness; the past that refuses to be pinned down always haunts the present. We are always undone with our past; it never really passes us.

Often considered the height of Wong Kar-wai's directorial career, *In the Mood for Love* (花样年华, 2000) is a film of nostalgia, memory, and unresolved love stories. It reconstructs a past time that might have been forgotten by contemporary Hong Kongers. The past, unknown to the young generation, is excavated, re-imagined, and reconstructed by way of mise-en-scène, music, costume, cinematography, atmosphere, or indeed, a 'mood' that is less an actual past than an illusive/elusive "past-ness" that history does not offer (Lee 2016: 379). The story takes place in a nostalgic reconstruction of a Shanghainese émigré community in Hong Kong in 1962. The original Chinese title of the film means "the age of blossoms," a phrase used to lament the passage of time, youth, and love. Through melancholic quoted intertitles at the beginning and the end of the film ("One can almost touch the past, but it slips away." "History is like a dusty pane of a window. One sees it but cannot clearly see through it," (*In the Mood for Love*, directed by Wong Kar-wai, 2000) etc.), the film creates the impression that time always eludes our comprehension.

In the Mood for Love has two main characters—Chow Mo-wan (Tony Leung) and Su Li-zhen (Maggie Cheung)—whose lives intersect as neighbours. Both betrayed by their spouses, they find solace in each other as daytime friends and secret lovers but end up quietly going their separate ways. With schizophrenic characters, fragmented temporal-spatial arrangement, cutting-edge visual stylisation, and mnemonic duels between nostalgia and historical amnesia, Wong offers his audience no satisfaction with predictable, if any, plots; instead, he puts them 'in the mood.' Mood, in this case, is a vicarious experience, which is stimulated mostly by the time-image.

Towards the end of *Cinema 1: The Movement-Image* and throughout *Cinema 2: The Time-Image*, Deleuze privileges the time-image over the movement-image, for he sees the movement-image as unable to be freed from representation or subjectivity because time or duration is always psychologically determined by events on the screen. On the contrary, the time-image enables time to break free from an event and therefore creates a more powerful cinema. In a film, any image that is not simply representing itself, suggests multiple states of being, and propels its audience to recognise, recollect, or even hallucinate could be a time-image.

In the case of *In the Mood for Love*, certain techniques, namely the *mirroring/doubling* of images (surfaces such as mirrors and glasses that refract and reflect) and characters (Chow Mo-wan and Su Li-zhen as doubles of each other's spouses) and the *repetition* of (non)events that disrupts and disorients the sense of time, are used to shift imperceptibly across the border between the present and the past/future, the actual and the virtual, the known and the unknown. With mise-en-scène, cinematography, and these editing techniques, the past and the future are shown to coexist, become indiscernible, and mutually inform the time-image.

In Wong's *In the Mood for Love*, we see the mirroring of various sorts, which is by no means coincidental. The film was inspired, in both form and content, by a 1972 Hong Kong short story, Liu Yichang's "Duidao" (*Tête-bêche*, or Intersection, literally meaning "upside down"), which later was expanded into a full-length novel in 1993. The novel's title, *Tête-bêche*, refers to stamps that are misprinted with the top of the one joining the bottom of the other. In other words, it carries the meaning of a double, or an upside-down reflection,[1] which becomes a trope throughout the film. Doubling or mirroring signals a dilemma, a constant pull between two identical but opposite entities that exhausts any possibility of progression. Characters are found in a trance-like state, misplaced in time and space, and unable to resituate themselves in the known spatiotemporal system.

In Wong's meticulously designed mise-en-scène, reflections and refractions shown by mirrors and glasses split and fragmentise the characters. In lieu of a Lacanian attentiveness of the self, the characters gain no identification with their own bodies; rarely do they look at, or are even aware of, their own reflections: they are meant to be looked at merely by the audience. Mirrors provide us with virtual images of actual entities, a series of temporal 'wormholes,' a short cut from the present to the past and

---

[1] See also Luk (2005).

future. In the film narrative, two characters that resemble each other can be seen as mirror-doubles. Su and Chow's lives have a double existence, as they play their own roles as betrayed spouses and, at the same time, they role-play as the doubles of each other's spouses when they imagine their own illicit affair. Resembling actual mirror shots, doubling in film disrupts the otherwise linear flow of time in a film: it creates a temporal short circuit. The doubling forms a complex and highly ambiguous story. On one hand, Su and Chow are the main characters of the love story, making them seem more 'real' to the audience. On the other hand, their spouses' infidelity is what drives them together before they had any mutual romantic feelings, which, in a sense, makes their relationship appears less real than that of their unfaithful spouses. Who, then, is the virtual image and who is the actual entity? When one image alludes to another, causing a chain of meanings in both images, the time-image occurs.

An example of a mirror/time-image is when Su Li-zhen, who has discovered her husband's betrayal, confronts Mrs. Chow with suspicion. The first shot of this sequence, which we see in extreme haziness, is Mrs. Chow walking towards the door. As she opens the door, we see a view of her back in a narrow mirror at the right side of the frame, splintering the interior space. The camera then cuts to a frontal shot of Su, directly facing Mrs. Chow as well as the audience. When Su talks to Mrs. Chow, instead of a series of shots and reverse shots of the two, the conversation is concluded by a single long take of Su's face, hesitant and restrained, interacting with Mrs. Chow (we only hear her voice, in fact, we never see her face in the entire film). In the last shot of this sequence, Su's frontal shot is cut into a side shot as Mrs. Chow closes the door. This sequence of shots is itself mirror like. It stimulates anticipation among the audience but refuses to give a clear answer (of what happened, how do their spouses look, etc.). Instead, it restricts the audience to look for answers only within Su. The duration of Su's close-up indicates the dual existence of an external physical time, during which she converses with Mrs. Chow, as well as an internal mental time, for with the absence of Mrs. Chow in this sequence, it indeed looks less like a conversation than Su's own internal monologue, confronting her own doubt. This paradox sets the overall tone of an irresolvable melancholic longing, for one cannot resolve situations where the characters are presented as a shadow of another person.

Another technique used to produce time-image is the repetition of banal, mechanical gestures: this includes not only actions of the protagonists but also repetitive scenes of the neighbours playing mahjong and Mrs. Chow calling her lover at her workplace. One example is the sequence of Chow and Su's staircase encounter. Often known as Wong Kar-wai's signature montage footage, it uses slow motion that matches the rhythm of the background music, recording Su's and Chow's walking up and down the stairs. The two encounters/miss-encounters resemble each other but lead to different outcomes. Time is made visible, as if it were sliding along the surface of Su's dress. Yet these chance encounters look less like designed plots than what Deleuze calls "*pure optical ... situations*" (1986: 120 [emphasis in original])[2] where sequences are used not for the sake of the progression of narrative but for the sake of the act of looking itself. This sequence is "dislocated in [narrative] time" (Deleuze 1989: 5), rendered completely random, and not subjected to the rules of action and response, cause and effect. It self-reflexively and self-consciously alienates the audience from the dominant cinematic experience of narrative time. The scene where Su restlessly runs up and down the stairs is strikingly similar to the searching scene in Hitchcock's *Vertigo*, which is, as Deleuze observes, an inversion of the spectacle and the spectator. The character Scottie becomes a kind of spectator, shifting, running, and becoming animated in vain while no longer subject to the rules of response and action. Instead of being engaged in an action, he becomes prey to a vision, pursuing it or being pursued by it (Deleuze 1989: 3). Similarly, the pure optical situations in *In the Mood for Love* engage the senses of the audience directly with time and thought. This is precisely what the time-image does: it aims to make time or thought sensible (Moulard-Leonard 2008: 109) and to free time from its subordination to movement.

These events, or to be more precise, 'non-events,' do not happen in a particular time or space. Space here is more affective than perceptive: Chow and Su may or may not meet, it may or may not rain, the indecisive Su may or may not accept Chow's ship ticket. Space in this case is not where events take place. Instead, it is a site where the virtual—the memory and the desire (the memory of their last encounter and the desire for the next encounter)—intersects with the present. These moments of encounter, in its multiplicity, is a fluid amalgamation of past, present, and future. Compared to the dominant Hollywood narrative style, *In the Mood for Love* is somewhat anti-narrative: it depicts a world that is often "dispersive, elliptical, errant or wavering, working in blocs, with

---

[2] See Gheorghe (2013: 93–99).

deliberately weak connections and floating events" (Deleuze 1989: 1). With the traditional film narrative, the audience look for the story *behind* the event, always searching for the internal logic and connection between events. Wong's audience, on the other hand, is drawn to the event itself, the event in its entirety and imperfection. In the sequences of (non)events in *In the Mood for Love*, there is no actual story behind the event, the story *is* the event. In this manner, the audience never knows which aspect of an image will be selected for reworking later in the film. Each image becomes suffused with uncertain relationships. What is in these shots is a direct depiction and imaging of time (a time-image), a mutation of the dominant structure of the cinematic time, which is often perceived through movements and the progression of narratives (movement-image). The juxtaposition of the time-image and the movement-image creates a tension between two distinct ways of perceiving time and in turn highlights and intensifies the audience's cinematic experience of time.

As suggested by the foregoing observations, Wong Kar-wai's *In the Mood for Love* belongs to the cinema of time-image. By using doubling/mirror-images and repetitive pure optical situations, it gives the audience an impression of something virtual or of a dream and put its audience in a 'mood,' a state of delusive wondering that is provoked during the film and prolonged into their own reality. Compared to the lucid, ready-to-consume Hong Kong commercial films, in *In the Mood for Love* Wong not only refuses to employ the conventional conception of time to advance the film's plot, he makes time itself enigmatic: the narrative stalls and relapses, the characters hallucinate in delusive rehearsals of life (with no actual actions or solutions), time is spatialised (the only way to tell it is a different day is by the different pattern of Su's dress), and space is temporalised (Chow tries to preserve his memories for Su by telling his secrets to a tree). Wong's films urge their audience to re-conceptualise time, both in the cinematic world and in their own.

In keeping with Wong Kar-wai's other films, which interrupt and invert the sensory motor schemata of the movement-image, *2046* (2004), an informal sequel of *In the Mood for Love*, continues to ponder over the uncanniness of lost time and the recollection of memories with the use of time-image, more specifically, what Deleuze defines as the "crystal-image" (Deleuze 1989: 69).

In this genre-bending film, Chow Mo-wan, the protagonist from *In the Mood for Love*, indulges in a hedonistic lifestyle in the aftermath of his unconsummated romance with Su Li-zhen. *2046* consists of four story arcs, frequently jumping from one temporality to another. The 'main' plots can roughly be summarised as follows: In 1966 in Hong Kong, Chow, a reporter turned erotica author, idles away his free time in night-clubs and banquets. He moves into an apartment building and settles in room 2047, while waiting for room 2046—the room number of the apartment where he used to spend time with Su—to be refurbished by his landlord. He gets involved in an affair with his neighbour Bai Ling (Zhang Ziyi) and develops a vaguely romantic relationship with his land-lord's daughter Miss Wong (Faye Wong). The eccentric and shy Wong is burdened by her unsuccessful relationship with a Japanese man (Takuya Kimula). To encourage her, Chow starts to write a science fiction novel called *2047* and reinvents Wong's love story with her boyfriend. The story takes place on a train departing from the year 2046, a year which attracts time-travellers who seek to reclaim their lost memories; 2046 also refers to a temporal space where everything remains the same in eternity. While writing the novel, he starts to project his own memories and desires into his story.

In *2046*, the shifting narrative is accompanied by juxtapositions of scenes of different temporalities, rendering reality indiscernible from memories and fictions/fantasies. At the outset of *2046*, for instance, we see a futuristic world with lavish skyscrapers and smoothly running trains with the voice-over of Tak (Takuya Kimula), the protagonist of Chow's novel *2047*. Later we hear Chow's monologue. The scene is then cut to a flashback sequence, when Chow left Singapore and saw his former love interest Black Spider (Gong Li) for the last time. The next shot again takes us to the fictional world of *2047*. Immediately after, as Chow recalls his traveling back to Hong Kong from Singapore, the shot is cut to a realistic, documentary-like black-and-white footage of Hong Kong's 1966 riot. A compound of wonder and confusion, the story gradually unfolds itself. This plot arrangement aims at disorienting the audience, who tend to regard the futuristic world as ontologically stable as the film's period setting (Bettinson 2009: 169).

In addition to the self-consciously misdirecting narrative ploys, the reappearing image of a tree hole—a visual clue left at the end of *In the Mood for Love*—also puts the working of time into question. This cen-tral trope can be understood to refer to what Deleuze defines as the "crystal-image." As Deleuze writes: "What constitutes the crystal-image is the most fundamental operation of time: since the past is not consti-tuted after the present that it was but at the same time, time has to split itself in two at each moment as present and past. ... Time consists of this split, and it is this, time, that we see in the crystal" (Rushton 2012: 83, quoting from Deleuze 1989: 81).

Deleuze is concerned with the ways in which time is conceptualised as memory. If we consider memory, that is, the act of remembering the past, as a flow of consciousness, where does the past begin and the present end? When does one start to recollect? Deleuze's answer is a simple yet enigmatic one: it is extremely difficult to demarcate the two, for they coexist in every moment of one's life. Deleuze's conclusion is indebted to Bergson's understanding of the *déjà vu*. In this particular phenomenon, we are in a place or situation that is uncannily familiar, as if there were a flow of memory from our past that merges with the present. We remember things from the past instantly when things happen in our present. Therefore, what we experience in the present unavoidably involves recollecting the past. Memory is formed not in the past but in the very present, which, according to Bergson (paraphrased by Deleuze), is "the recollection of the present, contemporaneous with the present itself" (Deleuze 1989: 79). Every moment can be split into an actual present and a virtual past, one carrying the other along with it (Rushton 2012: 82). The crystal-image entails this splitting nature of time. It captures the indiscernibility of the actual and the virtual and the fleeting moment of their interactions. Noticeably, this indiscernibility, Deleuze emphasises, is an "objective illusion" (Deleuze 1989: 69). The intermingling of the present and the past revealed by the crystal-image is not an illusion in the individual mind but an objective fact: it is a process in which the virtual becomes actual and vice versa. In other words, it is "the objective characteristic of certain existing images which are by nature double" (ibid.). Thus, the crystal-image is different from a flashback, in which a memory is summoned by a subject, usually to respond to or reveal something in the present. The crystal-image is formed when the past and present become unattributable. It is the image of a condition in which the actual and the virtual are entirely reversible.

There are numerous examples of the crystal-image in Wong Kar-wai's films: the image of the clock in *Days of Being Wild*, mirror-images in *In the Mood for Love*, and perhaps the most enigmatic among all, the tree-hole image in *2046*. In the beginning sequence of *2046*, Tak whispers into a 'tree hole,' or more precisely, a tree-hole-like mechanism on the wall of the train leaving 2046. This 'tree hole,' which appears to be something resembling a large phonograph horn built in a wall, becomes identifiable when Tak utters:

> I heard in the old days, when people had secrets they didn't want to share, they'd climb a mountain, find a tree, carve a hole in it, whisper the secret into the hole and cover it up with mud. That way, nobody else would ever learn the secret. ... (*2046*, directed by Wong Kar-wai, 2004)

This scene corresponds to the end of *In the Mood for Love*, where Chow, having lost Su, whispers his secret to a tree hole in Angkor Wat. This 'secret,' concealed from the audience in *In the Mood for Love*, is revealed in *2046* when Tak, Chow's fictional incarnation, asks his lover: "Will you go with me?" (ibid.). It is the same question that Chow asked Su, but he never gets a definite answer. Over time, this unanswered question has become a secret to Chow, a haunting question that ultimately refuses to be answered, a perpetual wondering. The sense of wondering is prolonged into *2046*, splitting its myriad memories, fantasies, and ongoing affairs into a state of uncertainty and ambiguity. This is why, as mentioned earlier, the fictional story of *2047*, Chow's way of reconstructing his memories, seems as real as his on-and-off relationship with Bai Ling. Here lies another characteristic of the time-image: to see time in its pure state is to open up time and to deny closure. The secret, which is concealed in a tree hole but constantly resurfacing, suggests the immanent splitting and opening up of time and thus constructs a grand crystal in *2046*.

THE UNKNOWN FUTURE: *HAPPY TOGETHER*

In his conclusion to *Cinema 1*, Deleuze states that the cinema at the end of the Second World War is filled with *clichés* and he champions a "new image" (Deleuze 1986: 213). This new image has five characteristics: the dispersive situation, the deliberately weak links, the voyage form, the consciousness of clichés, and the condemnation of the plot (ibid., 210). This new image—the time-image—must emerge from the cliché, fully aware of its immanence, to produce new substances. Wong Kar-wai, a so-called Second Wave director, is fully aware of the clichés produced in (and the influx of clichés into) Hong Kong. Clichés pervade Wong's films, especially in his other critically acclaimed film *Happy Together* (春光乍泄, 1997). It is a film about the tumultuous relationship between two Hong Kong gay men, Ho Po-wing (Leslie Cheung) and Lai Yin-fai (Tony Leung) during their journey in Argentina. The film, taking up the clichéd road movie genre, is named after a song from an American rock band. It is an aggregation of universalised, commodified existence:

tango, cigarettes, polaroid photos. Deleuze describes these clichés as "floating images," which circulate in the external world, but also penetrate our internal world (ibid., 210). Not only in *Happy Together* but also in his other films such as *In the Mood for Love* and *Chungking Express*, the clichéd images create a sense of endless floating and circulating, as Ho says repeatedly to Lai: "Let's start over" (quoted in Rushton 2012: 148). The circulation of clichés and the proliferation of meanings resemble the working of time in Wong's films. With the breakdown of the sensory motor system, time is subjected to expansion and extraction and collapses over and over again. How does this, then, as Deleuze questions the post-war cinema, "upsurge [a] new thinking image" (Deleuze 1986: 215)? What are the politics and consequences of this new image? We propose that through the time-image in his films, Wong creates a space for the past to negotiate with the present and brings about a constant state of becoming and renewal.

The 'Hong Kong Second New Wave' directors, including Wong, emerged in the 1990s, a period of uncertainty and anxiety resulting from China's impending takeover of Hong Kong. Wong also explores such topics, but in his idiosyncratic, elusive style. For example, 2046 is a hotel room number, and also a year, referring to the putative lifespan of Hong Kong's capitalist system. Hong Kong returned to China in year 1997, under the then Chinese leader Deng Xiaoping's promise that the capitalist system in Hong Kong would not be changed in the next fifty years. With this statement, he hoped to convey a sense of the determination of the Chinese Communist Party not to terminate Hong Kong's social system. 2046 is the year before the capitalist system in Hong Kong would expire. Hence the pun in the title of the film *Happy Together* also refers to the uncertain future of Hong Kong's imminent return to mainland China in 1997. The first shot of the film is a shot of two Hong Kong passports. The passports indicate the 'British Nationality' of the two men. Yet, later that year, Hong Kong will revert to China on July 1, 1997. Key themes of the film include a quest and a journey. It compels us to think: What will happen to Hong Kong residents afterwards? Will their British nationality expire? Toward the end of the film, Lai stops by Taipei before he returns to Hong Kong after the long trip in South America. In his hotel room, China Central Television broadcasts the news of the death of China's paramount leader Deng Xiaoping in early 1997: this is a subtle allusion to Chinese geopolitics. In the last sequence, Lai takes the metro (MTR) in Taipei and travels into the future. The scene is filled with bright lighting, exhilarating accelerated tempo, and the movements of both the people and the train. The future is decidedly unknown, and yet the protagonist is on a fast train to the future. The film abruptly ends without a conclusion, suggesting that the future is unknown, open-ended, and yet to be lived. Personally, will Lai be happy together with his gay lover Ho Po-wing? Geopolitically, will China, Hong Kong, and Taiwan be happy together in the unknown future? The film is open ended, opening up the issues of identity, nationality, citizenship, and colonialism in a rather personally felt manner.

In conclusion, Wong Kar-wai's entire oeuvre revolves around a continual working over the past, with the attempts of the characters to remember, replay, and resurrect the past. However, it is not a past to be closed or fixed. Instead, it is a past that is constantly undone, re-collected, and re-enacted. Wong's films reinvent the past with exemplary time-images and entail the openness of time. At the same time, they provide a new way of viewing Hong Kong's future. The conventional conceptualisation of the past, as pointed out by Svetlana Boym in *The Future of Nostalgia*, is an "extreme version of the eliminational model of progress ... a kind of tunnel vision of the road toward the future" (Boym 2002: 348). Wong creates in his films a future that is similar to what Boym calls a "reflective nostalgia," which "backtrack[s], slow[s] down, look[s] sideways, [and] meditate[s] on the journey itself" (ibid.) Wong's films consist of cinematic images that venture to offer approximations of the mysteries of temporality.

BIBLIOGRAPHY

2046. Directed by Wong Kar-wai. 2004. Hong Kong: Sony Pictures Home Entertainment, 2005. DVD.

As Tears Go By. Directed by Wong Kar-wai. 1988. Tartan DVD, 2005. DVD.

Benjamin, Walter. 1968. *Illuminations: Essays and Reflections*, trans. Harry Zohn. New York: Schocken.

Bettinson, Gary. 2009. Happy Together? Generic Hybridity in *2046* and *In the Mood for Love*. In *Puzzle Films: Complex Storytelling in Contemporary Cinema*, ed. Warren Buckland, 168–186. Chichester: Wiley Blackwell.

Bogue, Ronald. 2003. *Deleuze on Cinema*. London: Routledge.

Botz-Bornstein, Thorsten. 2008. *Films and Dreams: Tarkovsky, Bergman, Sokurov, Kubrick, and Wong Kar-wai*. Lexington: Lexington Books.

Boym, Svetlana. 2002. *The Future of Nostalgia*. New York: Basic Books.

Chunking Express. Directed by Wong Kar-wai. 1994. London: Artificial Eye, 2009. DVD.

Days of Being Wild. Directed by Wong Kar-wai. 1990. Tartan DVD, 2005. DVD.

Deleuze, Gilles. 1986. *Cinema 1: The Movement-Image*, trans. Hugh Tomlinson and Barbara Habberjam. Minneapolis: University of Minnesota Press.

———. 1989. *Cinema 2: The Time-Image*, trans. Hugh Tomlinson and Robert Galeta. Minneapolis: University of Minnesota Press.

Fallen Angels. Directed by Wong Kar-wai. 1995. London: Artificial Eye, 2012. DVD.

Flaxman, Gregory. 2002. *The Brain Is the Screen: Deleuze and the Philosophy of Cinema.* Minneapolis: University of Minnesota Press.

Goodman, Nelson. 1978. *Ways of worldmaking.* Vol. 51. Indianapolis: Hackett Publishing Company.

Gheorghe, Cezar. 2013. Bazin Meets Deleuze: The 'Fact-Image' and 'Pure Optical Situations' in Italian Neo-Realism. *Film & Media Studies* 1: 93–99.

In the Mood for Love. Directed by Wong Kar-wai. 2000. Tartan DVD, 2001. DVD.

Lee, Vivian P.Y. 2016. Infidelity and the Obscure Object of History. In *A Companion to Wong Kar-wai*, ed. Martha P. Nochimson, 378–396. Chichester: Wiley Blackwell.

Luk, Thomas Y.T. 2005. Novels into Film: Liu Yichang's Tête-Bêche and Wong Kar-wai's In the Mood for Love. In *Chinese-language Film: Historiography, Poetics, Politics.* eds. Sheldon H. Lu and Emilie Y. Yeh. 210–219. Honolulu: University of Hawaii Press.

Martin-Jones, David. 2006. *Deleuze, Cinema and National Identity: Narrative Time in National Contexts.* Edinburgh: Edinburgh University Press.

Martin-Jones, David, and William Brown (eds.). 2012. *Deleuze and Film.* Edinburgh: Edinburgh University Press.

Moulard-Leonard, Valentine. 2008. *Bergson-Deleuze Encounters: Transcendental Experience and the Thought of the Virtual.* Albany: State University of New York Press.

Powell, Anna. 2007. *Deleuze, Altered States and Film.* Edinburgh: Edinburgh University Press.

Rushton, Richard. 2012. *Cinema After Deleuze.* New York: Bloomsbury.

Wilson, Flannery. 2009. Viewing Sinophone Cinema Through a French Theoretical Lens: Wong Kar-wai's *In the Mood For Love* and *2046* and Deleuze's Cinema. *Modern Chinese Literature and Culture* 21 (1): 141–173.

CHAPTER 12

Suspense in the Cinema: Knowledge and Time

Hauke Lehmann

The appearing itself is the only true object of knowledge
—if it can, in fact, be an object.
(Jean-Luc Nancy, *The Surprise of the Event*: 166)

The relationship of knowledge to time is central to suspense in the cinema. Rather than just playing with the spectator's fear of the unknown 'out there,' suspense suggests something far more radical: what is unknown, and what remains inherently unknowable, is not the future itself, but *our arriving there,* our *affective encounter* with it—not in spite, but in the very light of everything we think we know about what is going to happen. This confrontation with our own affect as something eluding us is what fascinates. Thinking about fascination with regard to 'unknown time' brings up the fact that to conceive of time, we need a

---

I thank the editors, especially Sibylle Baumbach, for their helpful suggestions for improving this essay.

H. Lehmann (✉)
Freie, Universität Berlin, Berlin, Germany

S. Baumbach et al. (eds.), *The Fascination with Unknown Time,*
DOI 10.1007/978-3-319-66438-5_12

perspective on time. It is exactly this problem of perspectivity that will prove to be at stake in the concept of fascination.

David Fincher's *Gone Girl* (USA 2014) opens with an epistemological problem that hinges on the question of perspective: How can we know what another person is thinking or feeling? This problem pertains to the basic conflicts marking the social dimension of human existence: we live together, yet there is a fundamental gap between another person's access to the world and our own. As Jean-Luc Nancy writes in *Being Singular Plural*:

> The other origin [of the world] is incomparable or inassimilable, not because it is simply 'other' but because it is an origin and touch of meaning. Or rather, the alterity of the other is its originary contiguity with the 'proper' origin. You are absolutely strange, because the world begins *its turn with you*. (2000a: 6 [emphasis in original])

At the same time, this gap, this difference, is not the last word. Rather, as there is not *one* origin but many, the world is dissolved into a plurality of perspectives. Difference serves as a kind of relay between them, interfering precisely where we believe our most intimate core of self-congruence to be. Nancy again:

> It is no accident that we use the word 'intimacy' to designate a relation between several people more often than a relation to oneself. Our being-with, as being-many, is not at all accidental, and it is in no way the secondary and random dispersion of a primordial essence. It forms the proper and necessary status and consistence of originary alterity as such. (ibid., 12)

Thus, we are faced with a very basic tension between different perspectives, different "origins" of the world. This difference in perspective is not a mere, indifferent circumstance—on the contrary, it is expressed as the "strangeness of a singularity." According to Nancy, interest and hospitality, desire and disgust depend on this strangeness (ibid., 8). In this way, strangeness fuels a whole range of contradictory affective responses that might well be summarised under the heading 'fascination,' including extremes ranging from horror to curiosity. Nancy goes on:

> We find this alterity primarily and essentially intriguing. It intrigues us because it exposes the always-other origin, always inappropriable and always there, each and every time present as inimitable. This is why we are

primarily and essentially *curious* about the world and about ourselves. ...
(ibid., 19 [emphasis in original])

This is why I would argue that it is a fundamental difference in perspective that accounts for our curiosity as well as for our horror—in short, our being fascinated or thrilled—when we experience suspense in the cinema. Because suspense, as I argue, does nothing but manipulate our relationship with the intimate: that which marks our difference and at the same time relates us to one another. One important affective mode of this relationship can be aptly described as 'being fascinated.' Understood in this way, fascination sets in motion a peculiar push–pull dynamic (oscillating between attraction and repulsion) that can provide the basis for a temporal composition. Fascination, like suspense, is a state where subject positions appear as not clearly separate and fixed. Rather, they emerge over the course of a temporal unfolding. It is just such a temporal composition that I want to explore in my analysis of *Gone Girl.*

"When I Think of My Wife, I Always Think of Her Head": *Gone Girl*

Gone Girl exemplifies the problem of perspectivity, paradigmatically, in the constellation of marriage. Here is a brief summary of the plot, quoted from the Internet: "On the occasion of his fifth wedding anniversary, Nick Dunne reports that his wife, Amy, has gone missing. Under pressure from the police and a growing media frenzy, Nick's portrait of a blissful union begins to crumble. Soon his lies, deceits, and strange behavior have everyone asking the same dark question: Did Nick Dunne kill his wife?" (Plot summary for *Gone Girl* on IMDb.)

As one can perhaps deduce from this description, it is the viewer whose perspective will be shifted and related to that of the different characters, developing, over the course of the film, an emotional interest into how things will evolve. What interests me is how this problem of perspectivity is staged in the film. As suggested by the very first scene of the movie, which will serve as a starting point for my analysis, the cinematic, audiovisual movement image cannot be pinned down to a single perspective. The film opens in darkness, with an electronic, dronelike score audible, higher and lower frequencies oscillating and intermingling. Then, a voice is heard: "When I think of my wife, I always think of

Fig. 12.1   The unfolding of an enigma (*Gone Girl*, directed by David Fincher, 2014. Los Angeles: Twentieth Century Fox, 2015. DVD, 0:00:27–0:00:48)

her head." From the darkness, the image appears as if called forth by the act of thinking invoked in the voice-over (Fig. 12.1), creating a parallel to the appearing/disappearing movement of the titles and suggesting a link to the title of the film, which indicates an act of leaving or vanishing. Already, several questions suggest themselves: Who is speaking? Who is seeing this image? Are these two perspectives of speaking and seeing identical? What status does this image have? Is it seen at all? Or is it a thought image, virtual, hypothetical, imaginary? Also, there is an echo on the voice-over—what is the relationship between sound and image here, especially the temporal relation?

The back of a woman's head becomes visible, caressed by a man's hand. On first glance, the image seems to duplicate the voice-over, or rather, to translate what is said into the realm of the visual—albeit with a slight twist, a slight extension: the act of thinking is complemented (transposed?—replaced?) by a caressing movement. On this first level, the image seems to fix the perspective of the voice-over: a husband, thinking about his wife—he might be identical to the male figure whose chest we can make out in the bottom of the frame. But if we listen closely to the voice-over, it does not describe a specific action but rather a more general idea that seems to resist translation into a single shot. And whom is he addressing? Is he talking to the audience? By extension, it is not clear whether the image describes an act of thinking at all. It could also be seen as a kind of response to the voice-over, imposing its perspective on it. Because of these several layers of uncertainty, the audiovisual image in its full sense can neither be pinned down to the auditory, nor reduced to the visual; it cannot be pinned down to one perspective. Instead, it opens up a *constellation* of perspectives, creating a kind of force field. Different senses—the reasoning voice, sight (implied by the gaze off-screen), and the touching hand—seem to compete for a privileged access behind the facial façade that is at the same time constructed as their opposing force. The image and the sound are not doing the same thing; they come together to create something that is more than their sum. They create an audiovisual image that is *realised in the perception of the spectators as an expressive figure of the scene's movement as a whole.* What is this audiovisual image that unfolds over the course of the scene? It is, I would say, the figure of an enigma.

The kind of audiovisual image I am referring to is not simply the 'figure' of a woman looking mysteriously and vaguely in the direction of the camera. It arises from an affective *relation* between the film and the audience, being realised in what Bergson calls *duration* (cf. Deleuze 1992: 8–11).[1] It is not what is represented on the screen but is the result of the audience's perceptual activity intertwining with the film's temporal composition of movement-images. It is the *gestalt*, the whole, the organisation of the cinematic world realised in the audience's act of perceiving. An enigma understood this way is not a thing but a relationship towards a thing: a way of perceiving the world. It is this way of perceiving the world, characterised by a certain temporality, a certain oscillating movement between being kept at a distance and being attracted that might be called 'being fascinated.' In this kind of understanding, figures such as the femme fatale or the head of Medusa—paradigmatic examples of fascination—are never just objects to be looked at. They always have to be thought of as opening up a relationship to the one who looks, as they themselves exert all kinds of forces (this is where the idea of the 'evil eye' becomes relevant; cf. Connor 1998). What I call an audiovisual or cinematic image is what emerges from an act of film-viewing as such a kind of sense: the "stylization" (Kappelhoff 2004: 169) of the audience's seeing and hearing according to a way of being-in-the-world. What we, as the audience, perceive are not objects (e.g., a woman's face), but *acts* of looking, hearing/talking, and maybe thinking, constructing the world and, in particular, the figure of a woman, in a certain manner. And what the film from the outset very subtly implies is that nothing about this manner is in any way self-evident.

---

[1] For an elaboration on Bergson's conception of time and how it is expanded in Deleuze's concept of the time-image, see Chap. 11 of this volume.

Fig. 12.2 The unfolding of an enigma (*Gone Girl*, directed by David Fincher, 2014. Los Angeles: Twentieth Century Fox, 2015. DVD, 0:00:27–0:00:48)

The scene continues, and this continuation makes explicit the latent discrepancy between the perspectives of image and sound. The voice-over states: "I picture cracking her lovely skull, unspooling her brains … trying to get answers," while the man's hand keeps caressing the woman's head. At the word "answers," the woman quite abruptly turns her head towards us, pushing the hand away and eventually fixing her gaze slightly off-screen. The voice-over continues, while the woman lets her head sink down slowly on the man's chest, all the while looking into the same direction (Fig. 12.2): "The primal questions of any marriage: what are you thinking? How are you feeling? What have we done to each other?"

With the first two questions, the voice-over's address changes from third person—"my wife"—to second person: "what are you thinking?" This device has the effect of suddenly aligning sound and image, with the question seemingly directed at this mute façade (the effect of a façade being accomplished mainly through an interplay between the cold, smooth texture of the image-space, the abrupt manner of the head's turning, and the scrutinising quality of the woman's gaze). The voice, however, retains its detached quality and thus does not 'enter' the image completely—we do not expect to find its actual source inside the frame. If this is an audiovisual image of marriage, husband and wife do not enter the same space-time; they meet only in thought or imagination, serving as projection screens for each other. The last question "What have we done to each other?"—now switching to first person plural—recasts this constellation of marriage into a straight conflict and into a temporal relationship of futility, of irreversibility, just as the image fades away. This sense of inescapable entanglement is augmented by the 'grammatical movement' of the scene, starting from first person singular and, step by step, casting the net wider to finally encompass every possible position of speaking.

If we can describe, in this way, the audiovisual image as a constellation of perspectives, then suspense is concerned with the *shifting* of perspectives. The opening scene of the film constructs the audiovisual image of an enigma, folded into a contradiction between tenderness and brutality. The brutality is delegated to the intimate perspective of a voice-over, an interior monologue, while tenderness is displayed on the surface of appearance. At the same time, the face of the woman as a paradigmatic marker of outward appearance remains opaque, inscrutable. This complex relationship between tenderness and brutality, display and opacity, will be unfolded over the course of the film by shifting between these different perspectives. This shifting also translates into the language of plot. Nick Dunne is suspected of murdering his wife, and for the first hour the film—operating like a police procedural—works hard to slowly raise suspicion in the spectator. Although, by completely focusing on Nick for the first hour, the film seems to take a perspective that is fairly close to him, this illusion of 'knowledge' is undermined several times (e.g., when Nick's secret girlfriend appears—hidden until then not from the perceptive Amy, as it turns out, but from the audience). It remains impossible for the spectator to be sure of Nick's 'true' feelings and thoughts—just as it was impossible for him (and us) to break through the façade of Amy's enigmatic expression in the prologue.

But, as Nancy reminds us, there is no missing the origin, as well as there is no appropriating or penetrating it:

> To reach the origin is not to miss it; it is to be properly exposed to it. … We reach it to the extent that we are in touch with *ourselves* and in touch with the rest of beings. We are in touch with ourselves insofar as we exist. Being in touch with ourselves is what makes us 'us,' and there is no secret to discover buried behind this very touching, behind the 'with' of coexistence. (Nancy 2000a: 13 [emphasis in original])

This view forces us to define the concept of knowledge in a way that is emphatically distinct from most of media theory: here, knowledge cannot mean access to a truth or a kind of information that could be stored or communicated as a content independent of form. Instead, it is the way the audience is positioned with regard to the film's events.

Knowledge in the sense relevant here is not generated and is not communicable; it is the communication itself, if communication is understood as being-in-communication, and not as sending or receiving a message. To paraphrase Nancy: what matters is not uncovering or penetrating a mysterious origin, a truth 'behind' appearances (although Nick may be trying to do exactly that) but precisely the *manner of appearance itself*, of being exposed to something or someone.

Knowledge is not simply 'there' in any abstract way and for itself; it is inevitably formed and transformed in real time. Knowledge, furthermore, can be related to the idea of intimacy, in the sense that the intimate in its very strangeness provokes the drive to knowledge, or curiosity. Nancy brings up this point when he writes:

> We are interested [in others] in the sense of being intrigued by the ever-renewed alterity of the origin and, if I may say so, in the sense of having an affair with it. (It is no accident that sexual curiosity is an exemplary figure of curiosity and is, in fact, more than just a figure of it.) (Nancy 2000a: 20)

In this context, the idea of "unspooling [Amy's] brains" (*Gone Girl*) seems to translate itself into the film's course of shifts and turns, surprises and reveals, its dramaturgy imagined as the unreeling of a spool. This way, the question of knowledge is transposed onto the dimension of *time*, or, to be more precise, onto the realm of experience unfolding over time. What I understand as knowledge cannot in any way be extracted from this process of unfolding—or unspooling, as the film would have it (it cannot be formulated as a propositional statement). As the mastermind behind several twists in the film's narrative, Amy operates as a stand-in for the figure of the director, manipulating the turn of events that is 'lived through' by the other characters and, not to forget, the audience (cf. Kasman 2014; Greifenstein and Lehmann 2013), while being herself threatened and manipulated by other characters in turn. In a way, even this distinction between characters and audience is undermined, as the audience is implicated into the process of transformation itself, while the characters are construed as spectators—spectators of TV shows, to be precise, or voyeurs of a life that unfolds just like a TV show. This portrayal raises the question of how suspense conceives of intimacy: is there intimacy without mediation, and, if this is not the case, how exactly is intimacy mediated?

On Cinema and Psychoanalysis

In the history of (Hollywood) cinema, the affinity between suspense and the intimate has found expression first and foremost under the rubric of psychoanalysis, standing as a cypher for everything that supposedly belongs to our deepest core of self but constantly escapes our attempts of controlling it—in short, the unconscious. It is no accident that the two oeuvres linked most closely to the elaboration of a poetics of suspense—Alfred Hitchcock's and Brian De Palma's—are also heavily and visibly preoccupied with problems and questions the cinema shares with, or takes over, from Freud and Lacan. Sometimes this is obvious to the extent that these problems can seem like mere surface effects, mere excuses for what is 'really' of interest in the films, especially ones as openly self-reflexive as *Vertigo* (Alfred Hitchcock, USA 1958), *Body Double* (Brian De Palma, USA 1984), or *Raising Cain* (De Palma, USA 1992). The notion that, as Freud put it, "*the ego is not master in its own house*" (Freud 2001: 142), can be generally interpreted as opening up a dimension beyond our selves—a dimension through which, in the cinema, the audience is implicated into a play of unconscious urges and desires, 'acted out' not only by actors, but always on the screen. This scenario is also how one might summarise one main strand of Hitchcock interpretation today (Wood 1989; Žižek 1992).

Cinema's interest in psychoanalysis lies with questions such as "Where do our thoughts and feelings come from?" or "What makes us 'us'?" And although Freud would look for the answer in models focused on the individual and the family (such as the Oedipus complex), there has been a substantial critique of his conception of subjectivity. As Nancy remarks, it is striking "that psychoanalysis ... represents a sort of paradoxical privatization of something the very law of which is 'relation' in every sense of the word" (2000a: 45). One strand of critique that is of particular interest to the questions this essay is concerned with directly refers to the role of the cinema.

In his essay "The Poor Man's Couch," Félix Guattari (2009) compares psychoanalysis (particularly the practice of the 'talking cure') and the cinema with regard to their different interactions with and mobilisations of the unconscious (cf. Kappelhoff 2004: 289ff.). He too emphasises the plurality and diversity of relations, of "libidinal charges" the spectator in the cinema is confronted with and opened up to. On the one hand, "[psychoanalysis] continues trotting out the same generalities about the individual and the family, while film is bound up with the whole social field and with history" (Guattari 2009: 257). On the other hand, although both psychoanalysis and the cinema aim at "transforming the mode of subjectivity of those who use [them]" (ibid., 259), they relate to the unconscious in very different ways: Psychoanalysis in its attempts of interpretation presupposes that the unconscious "is structured like a language" (ibid., 261) without taking into account how this structure was established in the first place, namely, in the complex interplay of societal forces. The cinema, Guattari claims, enables us to think of the unconscious as bound up in this interplay, to grasp the "polyvocity of its semiotic modes of expression" (ibid.), which are irreducible to the binary model of signifier/signified: "[Film's] montage of asignifying semiotic chains of intensities, movements, and multiplicities fundamentally tends to free it from the signifying grid that intervenes only at a second stage ..." (ibid., 263). Without a fixed relationship between signifier and signified, there are no stable conditions for what 'knowledge' can mean in any given instant:

> The semiotic components of film glide by each other without ever fixing or stabilizing themselves in a deep syntax of latent contents. ... Relational, emotive, sexual ... intensities ... are constantly transported there by heterogeneous 'traits of the matter of expression.' ... The codes intertwine without one ever succeeding in dominating the others; one passes, in a continual back and forth, from perceptive codes to denotative, musical, connotative, rhetorical, technological, economic, sociological codes, etc. ... (Guattari 2009: 263–264)

What complicates the function of knowledge with regard to film is not 'only' the dimension of time but rather the fact that film as a temporal composition realises itself in the spectator's bodily (affective, cognitive) activity of perceiving images. In this process of being embodied, audiovisual images mobilise the multiple different ways spectators relate to the world without ever settling on one for good; this takes us back to the problem demonstrated in the beginning of this essay. As argued in the analysis of *Gone Girl*'s opening scene, the audiovisual image cannot be pinned down to one perspective. It does not say anything, and, understood in a strict sense, it does not even show anything (cf. Guattari 2009: 264). What it *realises* as a spectator's affective experience are not objects but ways of perceiving the world, and this is why it can convey the experience of being fascinated. The image is only ever present in the entanglement of two acts of perception. It initiates and structures the creation and transformation of imaginary landscapes in which the spectator's sensations can find transitory shelters (cf. Kappelhoff 2004: 290). In turn, the spectator's unconscious "finds itself populated by cowboys and Indians, cops and robbers, Belmondos and Monroes ..." (Guattari 2009: 265).

To summarise: According to Guattari, the "social imaginary" modelled and activated by film is "irreducible to familialist and Oedipal models, even on those occasions when it puts itself deliberately at their service" (ibid., 257). This, I would like to argue, is exactly the case with *Gone Girl*: the film activates all the common tropes of cinema's affair with psychoanalysis: voyeurism, the fear of the castrating woman (the femme fatale), the commutations of innocence and guilt. But, similar to the films of De Palma, it inverts the equation: "if, for Hitchcock, cinema is a type of voyeurism, then for De Palma the reverse is true: that *voyeurism is a type of cinema*" (Dumas 2012: 80 [emphasis in original]), and, one should add, a type of television. Fincher's film is not concerned with simply adding another allegorical staging of the voyeur as the prototypical cinema viewer. Instead, it focuses on the question of how an image (a film image, a television image) is created, and what function voyeurism fulfils for this creation. As should be clear by now, by 'image' I do not mean the individual frame but the product of the entanglement between spectator and (TV) screen. Note that if one focuses on this primary entanglement, the difference between a film image and a television image becomes very pronounced. Steven Connor has distinguished the different types of screen with regard to theories of fascination from a psychoanalytical standpoint (Connor 1998: 15). Obviously—*Gone Girl* itself stages this very precisely, and often to comedic effect—there is a distinct manner in which the relationship between television and its viewers is constituted, a manner that can be clearly distinguished from the act of going to the cinema and watching a two-hour movie. In a sense, the whole dramaturgical premise of *Gone Girl* is based on this difference, this conflict of temporalities.

Therefore, the dimension beyond our selves opened up by renouncing the ego's control is nothing else but the audience's affective involvement into the film's temporal unfolding on the screen in the dark cinema. In the following, I describe the process of such an unfolding in detail, along with elaborating the question of affective involvement.

KNOWLEDGE AND SUSPENSE

Back to *Gone Girl*: after an hour of Nick acting suspiciously, implicating himself deeper and deeper—all the while testing the spectator's position regarding his believability, his capacity for violence and deceitfulness—it is revealed that Amy staged her kidnapping/murder as well as most of the clues pointing to Nick as the prime suspect. From this point on the formerly very controlled style of the film opens up in several directions, among them blatant media satire, comedy, and horror film. The film's culmination in this regard is Amy's gruesome murder of her former lover, Desi, whom she had used to hide from Nick. Here, the tension between tenderness and brutality is driven to its extreme. The scene is itself structured according to the mode of suspense: it is quite clear to the audience in which direction the scene will be going. But how do we know that? As I demonstrate in the following analysis, the audience's 'knowledge' in this case is nothing but the temporal gestalt of the scene, its rhythm and composition.

A basic principle of the mode of suspense is the establishing of relationships. The murder scene does this on at least three levels. On a first dramaturgical level, it performs the conclusion to a movement of stylistic and narrative escalation associated with Amy's narrative strand that begins roughly an hour into the film. On a second dramaturgical level, that of recurring patterns over the course of the film, it takes up the motif of Amy using sex as a trap for gullible men that was introduced before when Nick investigated Amy's former boyfriends. Thirdly, on the level of concrete staging (in the sense of mise-en-scène as well as microdramaturgy), it directly refers to and modulates a basic scenic constellation repeated and varied several times in *Basic Instinct* (Paul Verhoeven, USA 1992): sex turning into murder. For the sake of brevity, I leave aside these connections to focus instead on the unfolding of the scene itself.

The scene forms the direct conclusion of a period of preparation where Amy arranges everything for the murder she is about to commit to look like self-defence after being raped. It begins when Desi comes home one evening. Soon the two move to the bedroom where a contrast is developed between Amy's cold, decisive urge and Desi's shy attempts at kissing that transform into insecure hesitation in the face of her overwhelming determination. The calculated, dynamic quality of her movements is underlined by pointedly symmetrical compositions placing her

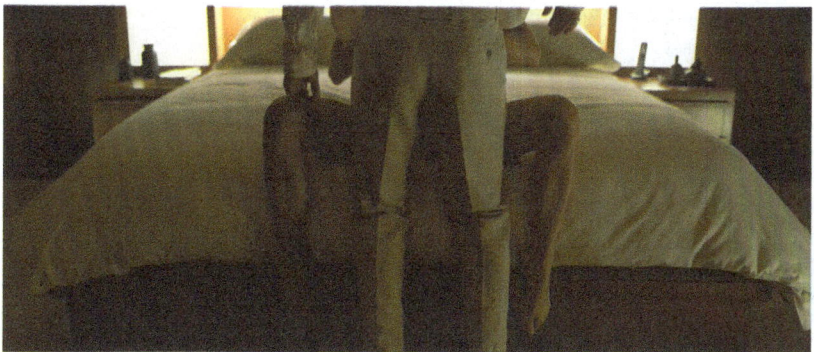

Fig. 12.3   The bedroom as a deadly trap (*Gone Girl*, directed by David Fincher, 2014. Los Angeles: Twentieth Century Fox, 2015. DVD, 2:04:08)

body in the centre of the frame (Fig. 12.3). Moreover, in these compositions and their montage, her body seems to merge with the whole room—filled with an abundance of rectangular forms—to build a trap that is about to spring. In this sense, it is only logical that the murder weapon (a box cutter) is already hidden in the bed, under a pillow. What is important in terms of the audience's involvement are not the represented actions but the duration of the scene as it is affectively experienced by the audience. The setting of the trap (i.e., the unfolding of the scene's basic conflict) is like winding up a mechanism aimed at the audience's sense of temporal balance.

At the same time, the staging of the scene works towards exposing Desi's body, transforming him into a vulnerable object. Here, the positioning of the bodies on the bed in combination with the framing is crucial: In contrast to Amy, who lies flat on the bed, Desi is moving into an almost upright posture while thrusting into her. The framing isolates him not only from Amy but also from the white bed, contrasting his figure with the yellows and browns of the background (Fig. 12.4).

In this moment of extreme vulnerability, shortly before (as it is implied) orgasm is reached, Amy fetches the box cutter from under a pillow (again, in a symmetrical composition) and slashes Desi's throat (Figs. 12.5, 12.6). This quick, sharp, precise movement from left to right interrupts the clumsy forward/backward movement of Desi's thrusting seemingly mere seconds before his climax. The parallel between sex and

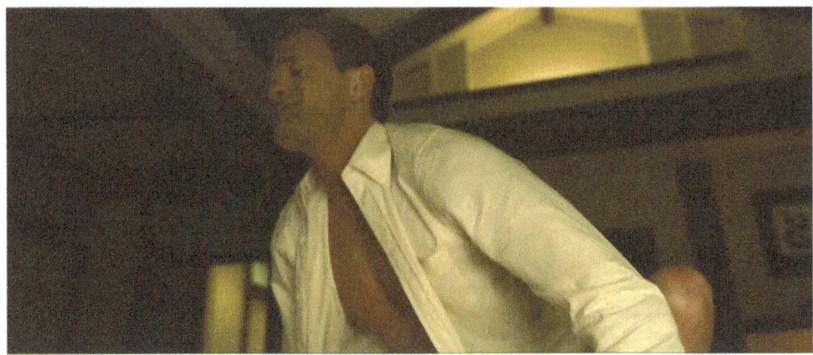

Fig. 12.4 Isolating a vulnerable body (*Gone Girl,* directed by David Fincher, 2014. Los Angeles: Twentieth Century Fox, 2015. DVD, 2:04:45)

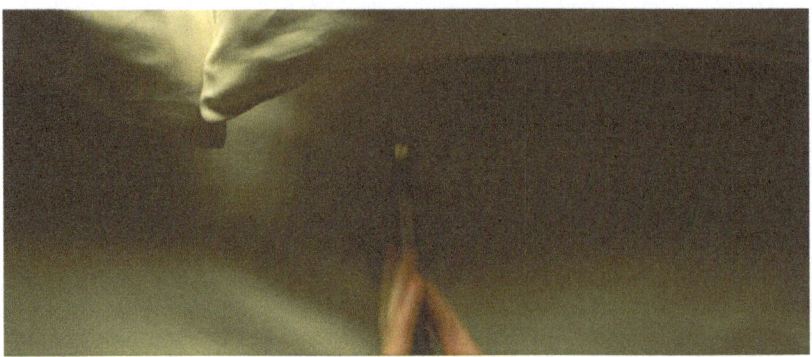

Fig. 12.5 The point of strongest conflict—the (premature) climax of the scene (*Gone Girl,* directed by David Fincher, 2014. Los Angeles: Twentieth Century Fox, 2015. DVD, 2:04:48–2:04:52)

murder is then further elaborated, the spilling of Desi's blood all over Amy and the bed substituting for his ejaculation. What remains is his facial expression of astonishment, caught wide-eyed in complete surprise.

'Knowledge' in this context designates the way in which the audience is affectively implicated into the developing conflict. The dimension of affective experience refers to the perception of qualities and intensities:

Fig. 12.6   Same as Fig 12.5

the clumsiness of Desi's gestures and movements in opposition to the exactness of Amy's movements coinciding with the geometric structure of the room, the slow crescendo of the rumbling drones on the soundtrack, etc. In this sense, knowledge is not so much a question of possessing this or that piece of 'information' (e.g., of having seen *Basic Instinct* or remembering the episode of Nick's investigation), but rather of affectively realising the setting of the trap as a time-image that aims at culminating and transforming itself into something new. The necessity of the scene's movement towards resolution does not refer to the actions represented on the screen; it is embodied in the audience's affective experience.

After the murder, Amy returns to Nick. The film ends with their reunion on national television, after which the introduction is taken up in a reprise. There, the arrangement from the beginning is repeated with some slight variations—Amy's mocking smile, the increased intensity of her movement, the recoiling of the man's hand—and a final, additional question: "What *will* we do?" By adding this question, the temporal structure of irreversibility, as introduced at the beginning, is opened up towards the future. In fact, both questions—"What have we done?" and "What will we do?"—could be combined into one: 'What will we have done?' And with that, a new tense would be introduced: namely, the future perfect, combining irreversibility and openness.

Anamorphosis of Time

To elaborate on this idea, we should return to the theory of suspense. The most famous example of suspense is Hitchcock's scenario of a bomb under the table: two people are talking, and at one point the audience is shown there is a bomb under the table and its timer shows it is going to explode in fifteen minutes. If the bomb is not shown but simply explodes without preparation, the audience will have fifteen seconds of surprise; if it is shown, however, Hitchcock claims, they have fifteen minutes of suspense (Truffaut 1984: 73).

From this point on, the audience is invested in the most banal conversation, because this conversation is no longer presented for its own sake. The audience's hopes and fears are being integrated into the film's process of staging, because they now have a privileged perspective on the proceedings—an intimate perspective that is split off from the formerly coherent context of the scene and which casts this scene in a new light, aiming at resolution and potentially undermining everything that happened before. Hence, suspense casts doubt not only on the audience's knowledge of the future but also, and to a much greater degree, on their presumed knowledge of the past and the present: without being able to exclude aberrant perspectives, how can we know we are on safe ground? This is the basic asymmetry in suspense: we as the audience can never be sure there will not be another 'under the table.'

With the introduction of the new perspective there are two orders of perception operating simultaneously in one and the same scene: the order of things happening above the table, and the order of things happening under the table. Because of this splitting, Pascal Bonitzer refers to suspense as an "anamorphosis of cinematographic time" (1992: 20), a distortion that requires the viewer to take on a new position. Anamorphosis introduces a new, aberrant perspective into a seemingly coherent context of perception. It implies two viewing positions contradicting each other.[2] This is what happens in suspense; but instead of being displaced in space (as with an anamorphic painting), the film viewer is displaced in time, is being made to imagine what is going to happen, to question what has happened before, and to impose this imagination and/or insecurity on the present.

---

[2] In this respect, the principle of anamorphosis can be compared to the kind of "double vision" analysed in Sect. The Speed of Apocalyptic Time of this volume with regard to science fiction writing.

The introduction of the bomb to the audience requires a change of perspective. In film, this change of perspective (as a change in knowledge) is inevitably carried out in real time, as the movement of a scene as a whole, stretched out across its duration. As the movement of the scene is realised in their perception, the audience is forced to realise this movement towards the new perspective. At the same time, they are privileged in being given this intimate insight, in being shown this secret that is about to turn the scene (or the whole film) on its head. Daniel Kasman elaborates on this movement of transformation with regard to *Gone Girl*'s macrostructure:

> The movie is consistent in one way up to the first twist, and then consistent in a second way afterwards. ... The prestige gloss of [Fincher's] images, the 'seriousness' of Affleck's performance, the superficial 'media critique' ..., all point to a surface-level graveness that the first part of the film tells us is sincere ... and the second part reveals is just as skin-deep as the characters Amy and Nick are playing. All is skin deep here; it's not just their marriage, it's the fabric of this movie's reality. ... We never see either participant in this marriage in any scene that's not mediated by some twist in narration or subjective view. In other words, we can't come to a conclusion about this couple like we can Bill and Alice in Kubrick's marriage film [*Eyes Wide Shut*], because all we see are 'spun' personalities. Instead, I find the film a very clever tweaking of genres of detection, crimes of passion, and indeed the kind of 'meditation of marriage' that the film seems to be; rather than be these things (about a mystery, about a hatred, about a marriage), it is [a] bizarre parody of those kinds of films, and as every parody does, reveals its cards by taking things farther than its sources of parody ever would. (Kasman 2014)

Kasman's description is very precise. The point I would like to emphasise is that, for the viewer, there is no direct relationship whatsoever to what the image presumably represents. Instead, the film constructs what Deleuze calls a "mental image" (Deleuze 1992: 203): an image that takes as its object neither an action nor an affect but a relationship, or, as Kasman writes, "some twist in narration or subjective view." Every image, instead of being self-sufficient, is distorted by or caught up with the "anamorphosis of cinematographic time" that is suspense.

If, because of its genre-slipping movement, Kasman calls *Gone Girl* a parody, I would add that this should not be understood as narrowing the film's ambitions; rather than making fun of certain genre conventions or trying to simulate the look of other films, the film seems to be concerned with the question of image creation in a quite general sense. The final question—"What will we do?"—is directed at the audience to make them imagine what could happen, make them produce images.

This activation of imagination is a direct consequence of the temporal anamorphosis of suspense that makes time as a principle of indetermination available as an object for reflection. Suspense enables the audience to reflect on their being suspended, on their being exposed to the arrival of the future. But this reflection is not a distant one, carried out by an already constituted subject. Instead, the audience themselves are caught up in the film's movement, and the medium of their reflection is the affective experience they are going through. As Nancy writes in a commentary on Deleuze's movement-image, significantly employing the future perfect tense: "Motion carries me elsewhere but the 'elsewhere' is not given beforehand: my coming will make of it the 'there' where *I will have come* from 'here'" (Nancy 2001: 28 [my emphasis]). If the viewer lacks a clearly defined position with regard to the movement-image, we cannot speak of him or her as an already defined subject. By undermining what we use to take for granted (the linear, chronological model of time), suspense also questions our position as perceiving subjects; this hinges on the fact that suspense makes explicit the inherent dimension of surprise in what Nancy calls an event:

> There is a rupture and a leap: rupture, not in the sense of a break with the already presupposed temporal continuum, but rupture as time itself, that is, as that which admits nothing presupposed. ... (Nancy 2000b: 170–171)

The event is not the result of a process but instead a *surprise*, an "interruption of the process" (ibid., 172) recasting everything that has happened before. This retroactive potency is what prevents any fixed positions from being constituted. If nothing presupposed is admitted, neither is our status as perceiving subjects. We are suspended in the overlapping of two temporal perspectives, unable to regain our footing in the present:

> When there is the event ..., it is the 'already' that leaps up, along with the 'not yet.' It leaps [over] every presented or presentable present, and this leap is the coming, or the pre-sence or *prae-sens* itself without a present. ...
> The tension or extension of the leap, that is, the spacing of time ...: this is

the surprise. The *Spanne* [referring to Heidegger, H.L.] is not surprising in that it comes to trouble or destabilize a subject that was there, but in its taking someone there *where he is not*, or insofar as it overtakes him, seizes him, paralyzes him *insofar as he is not there*. (Nancy 2000b: 171ff. [emphasis in original])

And thus, what is unknowable is not the future, not how things will turn out. There is no general 'how things will turn out,' no future 'as such,' because time is not thinkable without introducing a perspective on it. What fascination does is call into question any idea of a fixed perspective with regard to time.[3] Similar to suspense, it undermines the sense of a sure footing. Being "completely transfixed" (*Gone Girl*) in fascination (as Nick claims in an interview he was when he first met Amy) is precisely not to have a place but to be beside oneself. In fascination, there is not already a given 'I' that can say 'I feel.' This 'I' has to be produced in the entanglement between screen and viewer. So what is unknowable is what *I* will have made of 'how things will turn out,' because it is me as a viewer who has to perform the movement of the scene that is not given in advance. As Bill Schaffer writes, "What remains indeterminate even when the narrative outcome is certain … is not something in the film itself considered as an object kept at a distance, it is the film *in me*, as it lives in me. Ultimately, it is myself" (Schaffer 2000 [emphasis in original]). Suspense plays on this fundamental insecurity by slowing things down, by *giving us time* to imagine how things will turn out, because we as viewers are now intimately integrated into the film's process of unfolding—we ourselves have lost the stable ground beneath our feet. At the same time, there is an urge, a sense of inevitability, making us feel that we can do nothing but realise the film moving towards the future.

This confrontation with our own affect as something eluding us is what fascinates. Suspense is not a matter of forecasting an outcome but rather of negotiating for control over how we encounter the future. It refers not to what is represented in a given film but to the very process of the film's temporal unfolding. As an "anamorphosis of cinematographic time," suspense combines two temporal perspectives that seem mutually exclusive: the "already" and the "not yet."

---

[3] For the description of a similar destabilization of a spectatorial perspective, with regard to (deep) historical time, see Sect. Ways of Time-Making of this volume.

This is also the reason why we can experience suspense again and again, even if we already 'know' a particular film. In the same manner, *Gone Girl* elicits a fascination that can be experienced again and again, with each viewing of the film—a fascination which does not lose any of its attraction from information about the plot or the ending. It is not the film that changes but the way the film is constituted by ourselves as an object of feeling. Suspense is thus not about the simple dichotomy between knowledge and ignorance but rather about the unknowable dimension of time as it is lived by us while watching a film: it is about transforming our way of relating to ourselves as perceiving subjects. This, finally, is the reason for the disturbing and fascinating quality of suspense: it relates us to our own way of feeling, letting us watch and enjoy ourselves being thrilled. Because what is unknowable is not something out there but ultimately our own affect, our own arriving in the future that is not our own.

BIBLIOGRAPHY

Bonitzer, Pascal. 1992. Hitchcockian Suspense. In *Everything You Always Wanted to Know About Lacan (But Were Afraid to Ask Hitchcock)*, ed. Slavoj Žižek, 15–30. London: Verso.

Connor, Steven. 1998. Fascination, the Skin and the Screen. *Critical Quarterly* 40 (1): 9–24.

Deleuze, Gilles. [1983] 1992. *Cinema 1: The Movement-Image*, trans. Hugh Tomlinson and Barbara Habberjam. London: Continuum.

Dumas, Chris. 2012. *Un-American Psycho: Brian De Palma and the Political Invisible*. Chicago: Intellect.

Freud, Sigmund. [1917] 2001. A Difficulty in the Path of Psycho-Analysis. In *Complete Psychological Works*, vol. XVII (1917–1919): *An Infantile Neurosis and Other Works*, ed. The Institute of Psycho-Analysis and Angela Richards, 135–144. London: Vintage.

Gone Girl. Directed by David Fincher. 2014. Los Angeles: Twentieth Century Fox, 2015. DVD.

Greifenstein, Sarah, and Hauke Lehmann. 2013. Manipulation der Sinne im Modus des Suspense. *Cinema Yearbook* 58: 102–112.

Guattari, Félix. 2009. The Poor Man's Couch. In *Chaosophy: Texts and Interviews, 1972–1977*, trans. David L. Sweet, Jarred Becker, and Taylor Adkins, ed. Sylvère Lotringer, 257–267. Los Angeles: Semiotext(e).

Kappelhoff, Hermann. 2004. *Matrix der Gefühle: Das Kino, das Melodrama und das Theater der Empfindsamkeit*, vol. 8. Berlin: Vorwerk.

Kasman, Daniel. 2014. Dialogues: David Fincher's *Gone Girl*. *Mubi* October 1, 2014.   https://mubi.com/notebook/posts/dialogues-david-finchers-gone-girl. Accessed 27 July 2016.

Nancy, Jean-Luc. 2000a. Of Being Singular Plural. In *Being Singular Plural*, trans. Robert D. Richardson and Anne E. O'Byrne, 1–99. Palo Alto: Stanford University Press.

———. 2000b. The Surprise of the Event. In *Being Singular Plural*, trans. Robert D. Richardson and Anne E. O'Byrne, 159–176. Palo Alto: Stanford University Press.

———. 2001. *The Evidence of Film: Abbas Kiraostami*, trans. Christine Irizarry and Verena Andermatt Conley. Brussels: Yves Gevaert.

Schaffer, Bill. 2000. Cutting the Flow: Thinking Psycho. *Senses of Cinema* 6. http://www.sensesofcinema.com/2000/6/psycho/. Accessed 27 July 2016.

Truffaut, François. 1984. *Hitchcock*. New York: Simon & Schuster.

Twentieth Century Fox. Plot Summary for *Gone Girl*. *IMDb*. http://www.imdb.com/title/tt2267998/plotsummary. Accessed 27 July 2016.

Wood, Robin. 1989. *Hitchcock's Films Revisited*. New York: Columbia University Press.

Žižek, Slavoj (ed.). 1992. *Everything You Always Wanted to Know About Lacan (But Were Afraid to Ask Hitchcock)*. London: Verso.

CHAPTER 13

Futurology, Allegory, Time Travel: What Makes Science Fiction Fascinating

Kai Wiegandt

Literary critics have tended to regard science fiction as special interest literature. A look at bookstores, however, where shelf after shelf is filled with books about spaceships, aliens, unknown planets, and futuristic technologies suggests that these interests cannot be that special. Science fiction has long fascinated a large and diverse readership. Although most would intuit that the involvement of science fiction with futurity is fundamental to this fascination, it is less clear what exactly gives science fiction this appeal. I want to take a close look at how science fiction generates fascination through its treatment of time, beginning with the genre's evocation of the sublime. Turning to the example of Poul Anderson's *Tau Zero*, I discuss the relationship between science fiction and futurology on the one hand and science fiction and allegory on the other, and then argue for a shift in the genre's treatment of time and futurity in the 1960s.

In science fiction, future time enables the presentation of an unknown but not altogether implausible world that strikes the reader with a sense of wonder. Unknown places have always fascinated humans,

K. Wiegandt (✉)
Institut für Englische Philologie, Freie Universität Berlin, Berlin, Germany

© The Author(s) 2017
S. Baumbach et al. (eds.), *The Fascination with Unknown Time*,
DOI 10.1007/978-3-319-66438-5_13

but the successive discovery and mapping of uncharted territories since the Renaissance have left few places that science fiction could freely fantasise about in the manner of Jules Verne. The age of exploration had prepared the ground when around the beginning of the twentieth century H.G. Wells, in *The Time Machine* and other novels, introduced the crucial shift in science fiction from locating speculative literary worlds in unknown parts of the Earth to locating them in the future (Suvin 1979: 10, 208–221).

The sense of wonder often described as the emotional heart of science fiction has been specified by critics as the sublime: the reader is struck with terror while experiencing pleasure. According to Edmund Burke's classic definition, anything that is great, infinite, or unknown can be an object of the sublime (Burke 1987: 39–40, 57–74). Science fiction features both the natural sublime, for example, the rings of Saturn or vast uncharted territories, and the technological sublime, as in the case of spaceships or space colonies. Futuristic technologies of space travel strike the reader with awe for the sheer magnitude of spatial and temporal distances as well as the smallness of the Earth and the brevity of the lives of its inhabitants (see Mendelsohn 2003a: 3; Nye 1984: 14–15). The aesthetic category of the sublime is closely related to fascination—even more so than are the uncanny, the abject, or the fantastic. The sublime and fascination are emotional responses to objects or events that simultaneously elicit desire and repulsion. Both occupy the mind exclusively, leaving no room for concomitant emotions. Both leave the mind dumbfounded, although the elicited repulsion is less strong in the sublime than in fascination (Baumbach 2015: 152–153).

Although science fiction tells of the remote future, it does not present that future as disconnected from our present but as a continuous extrapolation of it. The sublime is presented as potentially real. Fredric Jameson has argued that fantasy narratives tend to short-circuit reality and utopia through magic, but science fiction affirms the reality principle at the cost of the pleasure principle of wish fulfilment.[1] The thought experiment, the 'what if' or "*novum*" (Suvin 1979: 63), is crucial to all science fiction and usually involves a particular scientific discovery

---

[1] Jameson's example for the reassertion of the reality principle is the Strugatski brothers' novel *Roadside Picnic*, the model for Tarkovsky's (quite different) film *Stalker* (see Jameson 2005: 74–76).

or invention.[2] Critics have observed that, in its reliance on speculation and thought experiments, science fiction is unique amongst other forms of fiction because here the idea is the hero (Mendelsohn 2003a: 4). The thought experiment both propels the plot and becomes a measure of the work's aesthetic quality (Suvin 1979: 15); it also establishes science fiction's proximity to futurology, the discipline that claims to scientifically predict the future. The proximity suggests that it is not the unbridgeable alterity of an unknown future that is fascinating in science fiction but the sense that its scenarios might with some plausibility come to pass, that we might be able to read the future in the present. Readers are fascinated by questions of science fiction's plausibility, accuracy, and different degrees of farsightedness, all of which can safely be determined only when the depicted future has become past.

To understand the nature of this fascination, it is useful to turn to Sibylle Baumbach's distinction of 'immediate' and 'prolonged' fascination elicited by texts. In the case of 'immediate' fascination, we are struck by a single event, image, or turn of phrase that arrests our attention. 'Prolonged' fascination envelops us more gradually as the narrative builds up tension by evoking stimuli that we cannot, at least for the moment, fit into a coherent concept. Both forms of fascination arise from a "disruption of familiar systems of signification, which alerts, astonishes and attracts until the unfamiliar is classified and the tension resolved" (Baumbach 2015: 29). This disruption, or cognitive disorientation, results from a failure to simultaneously conceptualise contradictory forces and leads to a breakdown of rational judgement (Baumbach 2015: 24–29). Science fiction's fascinating artificial worlds can be described as extensively elaborated, that is, structural rather than occasional, disruptions that force us to think of those worlds simultaneously as fictions of the present and as plausible future realities. Science fiction's disruptions of our sense of reality are structural, so they give rise to prolonged rather than immediate fascination. Although the reader gradually becomes used to the text's future world and learns to interpret that world as an extrapolation of the present, such extrapolation might lessen the sense of incompatibility between the present world and science

---

[2] Science fiction's reliance on thought experiments has led to the most influential alternative definition of the genre—speculative fiction.

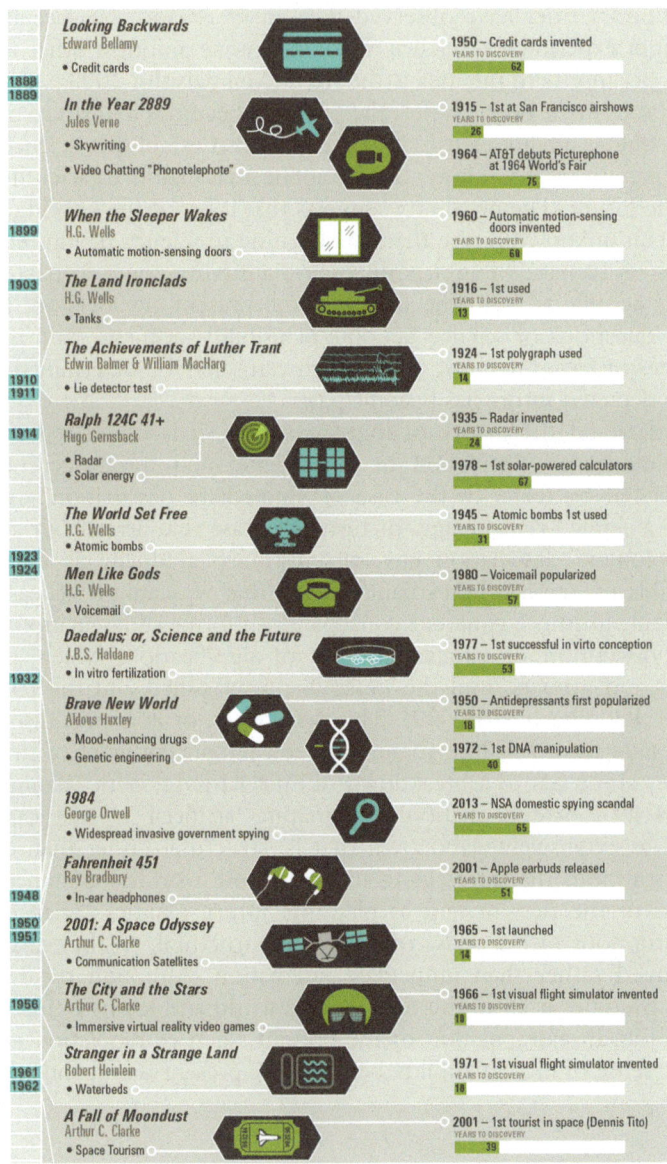

Fig. 13.1 Prediction or influence? Illustration from printerinks.com website ("A History of Books that Forecast the Future," *printerinks: Your Ink & Toners Shop.* Accessed January 13, 2016. http://www.printerinks.com/a-history-of-books-that-forecast-the-future.html)

fiction's possible future; however, it can also develop its own forms of fascination, as the following table (Fig. 13.1) illustrates.

The table shows early science fiction that predicted inventions in a future now past and illustrates the degree of science fiction novels' plausibility, accuracy, and farsightedness. For example, Huxley's *Brave New World* (1932) featured mood-enhancing drugs eighteen years before pharmaceutical companies popularised them and presented genetic engineering as a reality forty years before scientists succeeded in manipulating DNA. The table also raises the question of whether a particular work of science fiction predicted inventions or inspired their development. Did Ray Bradbury merely predict the invention of in-ear headphones in *Fahrenheit 451* or did he directly or indirectly influence Apple's engineers who designed them? The table can be read both ways. The question of influence is fascinating because it makes the reader think beyond the futurological idea that science fiction extrapolates the future and reflects it back to us, plunging us into paradox: science fiction that extrapolates the future might change that future and might thus undermine the necessary conditions of its extrapolation.

Critics have remarked that there are other reasons to doubt science fiction's capability to extrapolate the future. To begin with, only a small segment of science fiction called 'hard science fiction,' in which "a relationship to and knowledge of science and technology is central to the work," seems preoccupied with scientific extrapolation (Cramer 2003: 187). Authors of hard science fiction often have a background in the natural sciences and attempt to make their claims about future technology scientifically plausible, even if this means that their narratives are disrupted by physics lectures (Cramer 2003: 190).[3] The second and more serious objection is that there might be no reason to believe that any sort of science fiction can make valid predictions about the future at all. In Wells' *The Time Machine*, extrapolation of the social effects of the time machine are more important than the extrapolation of the scientific innovation itself. These effects can be "anticipated in a chronological future, but they cannot, scientifically speaking, be extrapolated.

---

[3] Hard science fiction goes back to problem-solving stories that focus on plot, such as Edgar Allan Poe's "A Descent into a Maelstrom" (1841), in which the hero uses scientific knowledge to escape a maelstrom. Pointedly put, there is a zone of indeterminacy between today's hard science fiction and the speculative scientific writing of, for example, research and development departments of big technology corporations.

By this token, futuristic anticipation reveals that extrapolating is a fictional device and ideological horizon" (Suvin 1979: 28) rather than scientific prediction. Especially from the late 1930s on, science fiction developed towards narratives where the disruptive invention eliciting a sense of wonder is only the starting point for an inquiry into the consequences of the invention (Mendelsohn 2003a: 4). Shifting from a model in which the future is deduced from the invention to a model in which the invention enables manifold equally possible developments, science fiction has moved from science into the sphere of anthropological and cosmological thought (Suvin 1979: 12). This development suggests that although thought experiment and invention have remained sources of fascination, science fiction must have tapped other sources to continue to fascinate its readers.

At least as fascinating as science fiction's tenuous futurological qualities, I argue, are the genre's possibilities of addressing the present through allegory, that is, through an extended metaphor whose details the reader gradually learns to interpret as corresponding to the details of some other system of relationships while simultaneously keeping their literal meaning in mind. When science fiction reflects present scenarios back to us in de-familiarised forms, it can fascinate us by trapping us in a double vision of conflicting timescapes, prompting us to decode analogies between future narrative and present experience, and to make sense of the differences that become visible against the background of these analogies, even while we continue to interpret its future world as just that: a possible world, actually existing in the future.

Science fiction stories typically combine the technique of extrapolation with the construction of analogies. These analogies problematise and often criticise the status quo through satire or grotesque elements. Science fiction narratives have developed such allegorical qualities to different extents and have done so to different degrees in certain eras. They belong in the tradition of tales such as Jonathan Swift's *Gulliver's Travels* or Voltaire's *Candide*, which similarly use allegory for critical purposes (Suvin 1979: 29–30). Reliance on allegory enables science fiction to present alternative worlds backed up by a myth of realism and rationalism and to imbue analogies with the present using that mythical authority: this implies that neither science nor the future is necessarily the main concern of science fiction. When in the 1950s serious research in telepathy was undertaken, a wave of books taking up the theme followed, and something similar occurred with nanotechnology and genetic engineering around the millennium. However, this does not necessarily mean that narratives extrapolating from these technologies chiefly explore their utopian or dystopian potential (James 2003: 228). Technology's principal function might be to furnish a science fiction narrative with new metaphors that reveal hitherto neglected aspects of the present while providing an illusion of realism and rationalism (Cramer 2003: 188). Even if a particular scientific problem or discovery is the motivation of a science fiction narrative and evokes the sublime, the narrative's main fascination might be created by the context of that problem and its analogies with the present. It is the combination of the hidden 'what ifs' and the initial thought experiment that prompts the reader's sense that the fictive world is both continuous and dissonant with the experienced world—a fascinating effect that is achieved not only by shifts of time, place, and technological scenery but also by style and lexical invention (Mendelsohn 2003a: 5).

To test this hypothesis, let me turn to a science fiction novel that does *not* seem to fit this claim. *Tau Zero*, published in 1970 by the physicist and writer Poul Anderson (1926–2001), is scrupulous in scientifically explaining its technological claims. It is considered a classic of hard science fiction, the branch of science fiction that is usually associated with futurological claims. The novel describes twenty-five men and twenty-five women setting out on the spaceship *Leonora Christine* to reach another star system after a (presumably nuclear) war has nearly extinguished mankind. The ship is powered by a Bussard engine, a type of propulsion that was proposed ten years before the novel's publication. The engine allows for speed approaching the speed of light, which means that the ship is subject to time dilation and relativity. The novel's title derives from the value of the time contraction factor tau (τ). The formula to calculate tau is given on one of its pages:

$$T = \sqrt{1 - v^2 / c^2}$$

where v is the ship's velocity and c the speed of light. At a given velocity, the duration experienced on Earth must be multiplied by tau to yield the duration experienced on board the ship. The closer the ship's velocity comes to light speed, the closer tau comes to zero, and the more time passes outside the ship in comparison with a given duration inside it (Poul Anderson, *Tau Zero*, 2006: 54–56, 90) (hereafter cited as *TZ*).

The crew's mission to find a new habitat for mankind comes at a price for everyone on board: during their five years on the ship, thirty-three years will pass on Earth. When the ship hits a nebula and the decelerator device is damaged, the device cannot be repaired because the engine must be kept running to provide radiation shielding. The crew are forced to continue at high speed, leaving Earth farther and farther behind (*TZ* 81–82). To reach an area where radiation is low enough to repair the decelerator, they accelerate (*TZ* 132),[4] with the effect that time dilation reaches extremes. After months they pass the mark when hundred years will have passed on Earth. Few of those on Earth they knew as children will still be alive. As they accelerate to shorten the time until they reach the next promising area for settlement, tau comes even closer to zero. Thousands, then even millions of years would have passed since their departure. The inconceivability of these durations translates into the willed helplessness of the prose: "They talked business for half an hour. (Centuries passed beyond the hull.)" (*TZ* 114).[5]

This synopsis makes clear that the novel's technological *novum*, the Bussard engine, is literally the engine of the plot. And yet, the novel's extrapolation of technological innovation, which leads to different timescapes inside and outside of the ship, is not primarily significant as a futurological vision of a world in which flight near light speed is possible. Instead, it fascinates us by teasing us into simultaneously thinking conflicting timescapes. Even more pertinent to my argument here, it creates the stage for a dilemma that has as much to do with the 1960s as with the future, which produces the aforementioned double vision. Although featuring lengthy scientific passages, much of the novel is concerned with the ship's constable Charles Raymont fighting the psychological effects of time dilation. The crew mourn those on Earth they will not see again and face the prospect of either suddenly dying by hitting a planet or living shorter-than-normal lives in the sensory deprivation of the spaceship.

---

[4] "There's no human difference between a million and a billion, or ten billion, light-years. The exile is the same" (*TZ* 132).

[5] For the problem of finding visual and verbal equivalents of extremely long stretches of time, see Chap. 1.2 of this volume.

The anti-aging cures available to them seem pointless under these conditions. They lose hope.

The situation on board *Leonora Christine* would have seemed strange to an American readership in 1970, but also strangely familiar. The threat of sudden annihilation had become real in 1962 with the Cuban missile crisis, when the Russians shipped nuclear missiles to allied Cuba, from where they could strike the east of the United States within minutes. After the threat was averted, fright and nervous expectation continued, as the public perception was of a world facing imminent destruction, and people suppressed their anticipation of radioactive doom from the skies (Broderick 2003: 48). *Tau Zero* begins *after* a nuclear war that has almost completely extinguished mankind. Crucially, the novel stages the possibility of imminent annihilation not by extrapolating another nuclear conflict but by way of the unstoppable Bussard engine that could cause a collision at any minute. When the crew escape near-annihilation on Earth only to find themselves in a similar situation in space, the novel is not concerned with the social consequences of Bussard engines but comments on life under nuclear threat in the late sixties. Nothing has changed after the Cuban missile crisis: the threat is still real.

Significantly, Anderson combines the threat of sudden annihilation with the availability of anti-aging cures and describes how single crew members choose not to undergo the anti-aging treatment because they do not want to live long under these conditions (*TZ* 114–116). He thereby captures a paradox plaguing the Western world during the Cold War: that the greatly extended lifespans of its citizens, showing in rapidly aging populations,[6] proved a dubious gift in a world under constant nuclear threat. As the crew of *Leonora Christine* waver between hope and despair, bouts of excitement and depression, and as readers compare these mood changes with their experiences of the 1960s, the fictional

---

[6]The life sciences have played a prominent part in science fiction, as progress in them is projected to affect longevity, intelligence and the brain, evolution, genetics, sexuality and reproduction, the environment and the biosphere (Slonczewski and Levy 2003: 175). Joan Slonczewski's *Daughter of Elysium* (1993) deals with some people who age, while others are engineered for near-immortality. In the last decade, speculations by scientists and inventors about the feasibility of doubled lifespans, new theories of aging as illness, and calculations of potential biological immortality of humans have surpassed the speculations of many fiction writers (see Knell and Weber 2009: 25–73; Hülswitt and Brinzanik 2010). For a discussion of science fiction's concern with extended lifespans (with a focus on Bruce Sterling's *Holy Fire*), see Mangum (2002: 69–82).

threat of fatal collision as well as the real threat of nuclear annihilation appear as dramatic versions of the general dilemma of mortality.[7]

The descriptions of festivities on board the *Leonora Christine* suggest that the imminent threat of death intensifies moments of exhilaration while simultaneously paralysing everyday life if hope is lost. Having out-lived their families on Earth and facing short lifespans, hope in the sur-vival of the species if not themselves emerges as central to the morale of the crew. Science cannot provide that hope. The ship's leading scientist, Elof Nilsson, can only present statistics about the unlikelihood of find-ing a suitable planet within their lifetime (*TZ* 120–122). Raymont has to maintain morale by assuming what he calls a position of hidden authority by instilling fear of punishment and irrational hope in the crew. Explicitly comparing this position to God's all-seeing eye, the secular-minded Raymont stands ambivalently between human efforts to assume God-like omnipotence and the belief that science—embodied by Nilsson—will never be able to replace religion.

The novel's staging of this tension speaks directly to the author's present. In 1969 the world had seen the first moon landing, sparking immense optimism about the future conquest of space. Millions had watched the landing on TV, itself a device that had been called "mere science fiction" by journalists only forty years earlier (Broderick 2003: 48). At this Promethean moment, nothing seemed beyond the human horizon. A realm that had been believed to belong exclusively to God seemed to have been conquered.

But had it really been conquered? In *Tau Zero*, technology fails twice: first the decelerator breaks, then technology fails as a source of meaning. Although disastrous for the crew, these failures provide the novel with new allegorical potential. Even if one does not read the claim into the novel that religion is an irreplaceable source of hope, the novel undeni-ably insists that a quasi-religious authority must be in place if the crew is to persevere. A critique of the myth of supreme scientific competence and galactic manifest destiny is implied. *Tau Zero*'s religious dimension is complemented by its plot's resonances with myths such as the *Odyssey*, Noah's Ark, the Flying Dutchman and *Moby-Dick*, all of which feature ships: as a toy at the mercy of greater forces, as lifeboat, as reified destiny

---

[7] A similar analogy is invited by two better known literary and filmic narratives: Kazuo Ishiguro's novel *Never Let Me Go* (2005) and Lars von Trier's film *Melancholia* (2011).

of perpetual travel, as symbol of the quest for the unattainable. Before anything else, however, the spaceship that cannot stop is an allegory of human mortality: thrown into existence, humans exist to meet their inescapable death.

In *Tau Zero*'s allegory of human mortality, the final twist consists in the fact that it is the universe that dies and is reborn after the ship has been flying at near light speed for years. If fascination is cognitive disorientation resulting from a failure to conceptualise conflicting forces in a coherent whole, the startling event of the universe's death and rebirth fascinates in multiple ways. It fascinates because the reader faces mutually exclusive planes of time: the linear time he knows and in which *Tau Zero*'s characters continue to age, and religious time breaking into secular time as it does in Christ's death and resurrection and in apocalypse (see Chap. 1.3). But the apocalyptic scenario does not merely belong to religious time but also to the supposedly secular 1960s and their constant nuclear threat: the secular and the religious, the linear and the circular bleed into each other, in this way disorienting, frightening, and fascinating American citizens as well as the readers of *Tau Zero* who, while reading about the universe's death and rebirth, read an allegory of the present as well as a narrative of a literally possible future. The discrepancy between time inside and time outside the ship now becomes readable as a meta-fictional pointer towards the novel's discrepant modes of signification: literally, as a tale about a possible future, and allegorically, as a tale about the actual present. Adding to the reader's entrapment in the cognitive impasse of fascination, religion and science emerge as equally valid conceptualisations of the universe's rebirth: some crew members experience the spectacle of the universe's rebirth as a second Genesis directed by God ("What is man, that he should outlive his god?") (*TZ* 171), but others consider Raymont their god-like hero who has led them out of the emptiness of space to the promised land (*TZ* 181–189).[8]

It is remarkable that Anderson addresses issues relating to the human lifespan, mortality, hope, and survival of the species on the religious and existential level rather than by way of the scientific innovations

---

[8]More or less direct invocation of religious questions and use of a religious register is common in science fiction where technological achievements suggest godlike power of the human. The tone surrounding in particular the question of immortality is often eschatological (Mendelsohn 2003b: 270).

of the Sixties. In 1959, Tofranil, the first anti-depressant drug, was released in the United States, with growing numbers of users (Shorter 2008: 60–64). Enovid, the first contraceptive pill, was released in 1960, and use of 'the pill' likewise became widespread during that decade (Coleman and Ganong 2014: 111–112). If Anderson had been interested only in extrapolating from contemporary science, he could easily have presented a scenario in which mood is pervasively controlled by drugs, as Aldous Huxley had done in his dystopian *Brave New World*. Drugs play only a minor role in *Tau Zero*, however, presumably because they would have interfered with the novel's interest in humans' own resources for coping with seemingly insurmountable challenges. Instead of proposing a merging of human and machine in cyborgs or the enhancement of humans through chemistry, the novel insists on what it posits as human authenticity. Raymont serves as the heroic embodiment of human ability, exhibiting more than anything else doggedness rather than scientific understanding or orthodox faith. It is this doggedness that leads him to making the right decisions, but the decisions themselves reflect Christian virtues. Assuming the position of a detached demigod, against all scientific reason Raymont instigates hope for a life after *Leonora Christine*, requests mutual charity amongst the shipmates, encourages them to become couples, and demands faithfulness to the ship's code of conduct.

Although *Tau Zero* avoids anti-depressants as an illicit and only temporarily effective solution to existential problems,[9] contraception plays an important role and invites comparisons with the author's present. The twenty-five women on board are requested to use birth control until a planet suitable for establishing a new civilisation is found. On board *Leonora Christine*, sex takes two forms. In combination with 'the pill,' it is a form of recreation and is recommended for therapeutic and psycho-hygienic reasons. Without the pill, sex becomes the dead-serious technical business of assuring the survival of the human race. Anderson witnessed the pill's coincidence with second-wave feminism and the sexual revolution of the 1960s, and likely saw a connection between the pill and changing views on sexuality and women's roles. The casual,

---

[9] The novel illustrates the illicit nature of drugs by featuring so-called dream boxes in which the crew can experience artificial but addictive rich sensory environments. Drugs or simulations of this sort are not a permanent solution.

even random sexuality on the spaceship, combined with the need to plan procreation, reads like a satire of alternative lifestyles explored during the Sixties.

Changing attitudes towards sex are only one component of these lifestyles obliquely reflected and commented on in *Tau Zero*. Although short-term relationships predominate on board the *Leonora Christine* while the mission continues as planned, after the accident marriage becomes fashionable, especially when the hundred-year mark has been reached—an illustration of the ways in which unknown experiences of time can revive rites that had been thought outdated (*TZ* 95–96, 101). The crew's composition of the world's ethnicities, apparently to recreate human civilisation in all its diversity, also links the old and archetypal with utopian times: a quaint reference to Noah's Ark, the ship's multi-ethnic composition reflects the growing multiculturalism and especially the Leftist ideal of multiculturalism in Western societies in the Sixties. When Raymont splits up with the ship's First Officer Ingrid Lindgren, he and Chi-Yuen Ai-Ling, a Chinese planetologist, begin a fulfilling long-term relationship. It would be far-fetched to think of John Lennon and Yoko Ono, but the famous couple of the late Sixties nevertheless embodies some of the values that are at stake in *Tau Zero*. Similar to many others, but much more visibly, Lennon and Ono celebrated multi-ethnic coexistence and questioned female and male roles.[10]

Let me summarise my argument up to this point. Anderson's novel allegorises mortality as an existential concern that sat heavily on people's minds in the 1960s, as well as other socio-historical features of the period. The novel's allegorical quality emerges as one of its greatest sources of fascination, as it puts the reader in a position of double vision: the world it presents can be read simultaneously as a possible future and as an allegory of the 1960s.

---

[10]Leonora Christine's crew is multicultural to an extreme degree, and the ship has a female First Officer, but in contrast to multiculturalism, feminism remains a fiction rather than a reality in *Tau Zero*. "Woman is the nigger of the world," Yoko Ono polemicised in 1968 in an interview with the women's magazine *Nova* (Lennon made a song of it, published in 1972). Her diagnosis can also be applied to the multi-ethnic space ship, wherein the First Officer Ingrid Lindgren, the highest ranking female, finds that the best she can do for the ship is to comfort and pleasure the male engineers, scientists, or constables when they feel dangerously low.

I now consider how time travel and allegory are related in *Tau Zero*, and how this relationship—which is symptomatic of the 1960s, as I show—fascinates the reader. To begin with, the novel's stress on allegory raises the question of whether time, whose passing the novel narrates, and the future in which it is set, are relevant only insofar as they allow for this allegory, or whether the passing of time is topical in itself. Most science fiction narratives begin in the remote future, and their narrators narrate in the past tense: the future is just another present that has recently become past, as in many narratives that are not science fiction. *Tau Zero* is narrated in the past tense, too, including the part before the Bussard engine is activated, which is already set in the future. Yet the time travel mechanism of the Bussard engine is used to make the reader witness the passing of time during the narrative and draws the reader's attention to time itself: time becomes topical. At the same time, however, *Tau Zero* focuses on allegory. How are its foci on time and on allegory compatible? Or are they related to one another? If so, how is this relationship itself related to fascination?

If measured by David Wittenberg's historical survey of time travel narratives *Time Travel*, *Tau Zero* is not untypical of the 1960s. Time travel narratives of the Sixties, Wittenberg argues, stand in marked contrast to time travel narratives written before Einstein's discovery of relativity such as H.G. Wells' *The Time Machine*. In these earlier narratives, the time machine mostly served as a motivation for deducing the future from the present, and especially for legitimising futurology with reference to dominant popular scientific explanations. The machine helped authors back up, with 'hard' futurological explanations, their often risky and taboo scenarios that fascinated readers. The explanations in question tended to be popular versions of evolutionary theory and mostly equated evolution with progress; they suggested that the utopian future was an outgrowth of the present (Wittenberg 2012: 30). Wittenberg argues that after Einstein's discoveries became popular, the relativity of time, time dilation, and compression become central in time travel narratives. Such narratives were no longer just a vehicle for utopian thought but formed an autonomous genre. Einstein opened up new plot possibilities: temporal dilation or reversal, physical access to one's own past or future, viewpoints encompassing all or other possible worlds, narcissistic or oedipal meetings with oneself and forefathers, etc. With these new possibilities, time travel narratives became increasingly self-aware of the recombination of narrative worlds and times. Time travel narratives today have developed into a literature about the forms and mechanisms of storytelling itself: they serve as a narratological laboratory (ibid. 31).

This is a compelling, although it seems incomplete, historical overview of time travel literature and its origins. Rather than crediting Einstein alone with triggering time travel's move away from utopian narrative, we must also take into account science fiction writers' awareness of the genre's history: the genre has now existed long enough for writers to look back and judge its different manifestations. In *Tau Zero*, the growing temporal discrepancy between people on Earth and on board *Leonora Christine* certainly testifies to an interest in Einstein's theory of relativity and its potential social effects in a future where near-light-speed travel might be possible. As this reading of the novel has shown, however, enabling the novel's allegory of the 1960s, including the paradox of longer lifespans coupled with a constant threat of nuclear annihilation, is an important, if not *the*, function of the Bussard engine as the material counterpart to Einstein's idea of the relativity of time.

In this concluding section of my chapter, I want to show that Anderson's allegory of the 1960s is a response not so much to Einstein's physics but to a growing scepticism amongst science fiction writers about the capabilities of science fiction itself. An important feature of that response is to make up for this scepticism by tapping new sources of fascination.

As Damien Broderick affirms, the beginning of the 1960s saw a crisis in science fiction writing. Many of the futurological claims of early science fiction had turned out to be wrong, and all these fictions said more about the time in which they were written than about the future that had then become present. Nothing, these negative examples suggested, can be more *passé* than past visions of the future. Science fiction writers became more aware of the fact that human situatedness in history, race, class, gender, and nationality co-determine imaginings of the future, and that the record of these co-determinations is the history of science fiction whose utopian futures had proven to be anthropomorphic projections of the authors' own societies and these societies' parochial concerns. Older science fiction came particularly into the crosshairs of the writers of the New Wave, the Sixties movement within science fiction following the examples of Truffaut's and Godard's *nouvelle vague* that set out to break with the genre's past, to explode the myth of supreme scientific competence and galactic manifest destiny (Broderick 2003: 49–52).[11] In 1961, Alfred Bester, a pioneer of the new science

---

[11] A critical attitude towards science became common in New Wave science fiction after J.G. Ballard (Broderick 2003: 52).

fiction, had launched the attack on conventional science fiction in *The Magazine of Fantasy and Science Fiction*: "The average quality of writing in the field today is extraordinarily low. ... Many practicing science fiction authors reveal themselves in their works as ... silly, childish people who have taken refuge in science fiction where they can establish their own arbitrary rules about reality to suit their own inadequacy" (Bester 2000: 400, 403). There was a strong sense that science fiction had to establish intellectually more vigorous standards for itself if it wanted to be taken seriously. That *Tau Zero* prioritised allegorising of the present over extrapolating the future can be read as a response to this call, but also as a response to scepticism concerning science fiction's capability to imagine the future.

Before I return to *Tau Zero*, let me explain the philosophical underpinning of the main problem science fiction writers faced, and which Stanislaw Lem's *Solaris* (1961) illustrated for a wide audience. The problem is located at the level of representation, as Fredric Jameson argues in *Archaeologies of the Future*:

> If there is nothing in the mind which was not already transmitted by the senses, according to the old empiricist motto, we are also generally inclined to think today that there is nothing in our possible representations which was not somehow already in our historical experience. The latter necessarily clothes all our imaginings, it furnishes the content for the expression and figuration of the most abstract thoughts, the most disembodied longings or premonitions. Indeed, that content is itself already ideological in the sense outlined above, it is always situated and drawn from the contextually concrete, even where (especially where) we attempt to project a vision absolutely independent of ourselves and a form of otherness as alien to our own background as possible. (2005: 170–171)

Solaris both explicitly discusses this empiricist argument and demonstrates it through its plot. A novelistic inquiry into the preconditions of the possibility of science fiction, it problematises earlier science fiction and was decisive in furthering the scepticism about science fiction's predictive power for technological progress that characterised Anderson's period (Broderick 2003: 49–52).[12] *Solaris* is concerned with human contact with an utterly alien intelligence, a planet-girdling, sentient ocean. For a century, researchers have tried to gain some understanding of this intelligent being, without success. Lem gives extensive accounts of the development of each new line of scientific inquiry and presents a miniature sociology of the scientists to demonstrate the systematic and emotional investment in humanity's effort to understand the foreign being (Jameson 2005: 108). One day the researcher Kelvin arrives on the space station hovering over the ocean to find that the crew have bombarded the ocean with gamma rays. From that moment on, strangers appear on board; the ocean seems to reconstruct them from the crew's memory traces, but it remains a mystery why the ocean does it. The novel ends with Kelvin's realisation that the other form of life cannot be understood unless the human observers shed their own humanity: "Where there are no men, there cannot be motives accessible to men. Before we can proceed with our research, either our own thoughts or their materialized forms must be destroyed" (Lem 1970: 134). There can be no "question of 'contact' between mankind and any non-human civilization" (ibid. 170). Lem's science fiction, Jameson argues, is designed to demonstrate its own absolute limits concerning the knowability of utterly other forms of life. For Lem, the only way to handle this dilemma was to illustrate it (Jameson 2005: 107–108). *Solaris*'s illustration of unknowability is as fascinating for the reader as the unknowable form of life is to the characters who fail to understand what to them is unthinkable: when the characters see and the reader reads about the foreign form of life, both experience what is conceptually impossible, a coming apart of thought and reality.

Lem did not take his scepticism to the last conclusion, however. *Solaris* is set in a distant future, and yet its scientist characters behave

---

[12] It goes without saying that also in the 1960s, most science fiction writers continued writing novels that did not exhibit such scepticism; just as most writers writing during Modernism did not write in a modernist vein. After all, the Sixties included important writers as different as Robert Heinlein, Frank Herbert, J.G. Ballard, and Philip K. Dick. Broderick suggests an additional source of the new scepticism when he explains that the late fifties and early sixties saw the arrival of the first generation of science fiction writers who were influenced by Modernism, which was itself characterised by epistemological scepticism (2003: 49).

much like twentieth-century men and women, rendering it easy for the reader to make sense of their motives and actions. According to Jameson, however, we must assume that future human beings will be quite different from us: the more different, the more remote from us in time, until they will be hardly intelligible to us at all. Lem's novel offers no justification of why human development, predominantly culturally but also physically, seems by and large to have stopped in the twentieth century. This problem of representation plagues all of science fiction: in a future world of radical difference, human beings would hardly be able to recognise themselves, whereas future worlds close to current realities are indistinguishable from the most common reformist or social-democratic proposals (Jameson 2005: 168).

Time travel devices might seem highly implausible, but in narrative terms, they offer a solution to the charted problem: a rationale for the fact that while the future world might be utterly alien, the protagonists are like you and me. *Tau Zero*'s use of the Bussard engine is a case in point. But this is not the only way in which Anderson's novel seeks a solution to the unknowability problem: its principal use of time travel for an elaborated allegory of the 1960s entails an extension of Lem's scepticism concerning the knowability of the future because it testifies to the realisation that the unknowability of the utterly other, as illustrated by Lem, also applies to the utterly distant future when time will have transformed the world and ourselves in ways that are hardly imaginable. Lem turned the science fiction novel into a philosophical tale about the unknowability of the radically other; in *Tau Zero*, Anderson goes beyond unknowability by focusing on what is knowable: the present.

Realising that limits to imagining the future limit possibilities of fascinating (future) readers with forecasts of alternative worlds, Anderson's allegorical science fiction generates its own form of fascination. Making the reader see what happens in the novel simultaneously as possible future and actual present, religion and science, awful and awe-inspiring, attractive and repulsive, the novel thwarts the reader's conceptualisation of information into a coherent whole. In narrating the possibility of accelerating the passing of time through time travel, the novel makes us realise the impossibility of representing the remote future: this is how *Tau Zero*'s stress on time and its allegory of the present depend on each other. If futurological science fiction runs the risk of becoming literally and metaphorically *passé* as years pass, science fiction as historical allegory can fascinate even future readers with its double vision, as I hope to have shown. It is not least in this sense that *Tau Zero* elicits prolonged fascination.

BIBLIOGRAPHY

Anderson, Poul. 2006. *Tau Zero*. London: Gollancz.

Baumbach, Sibylle. 2015. *Literature and Fascination*. Basingstoke: Palgrave Macmillan.

Bester, Alfred. 2000 [1961]. *A Diatribe Against Science Fiction*. In: *Redemolished*. New York: ibooks.

Brinzanik, Roman, and Tobias Hülswitt (eds.). 2010. *Werden wir ewig leben? Gespräche über die Zukunft von Mensch und Technologie*. Berlin: Suhrkamp.

Broderick, Damien. 2003. New Wave and Backwash: 1960–1980. In *The Cambridge Companion to Science Fiction*, eds. Edward James and Farah Mendelsohn, 48–63. Cambridge: Cambridge University Press.

Burke, Edmund. 1987. *A Philosophical Inquiry into the Origin of Our Ideas of the Sublime and Beautiful*. Oxford: Blackwell.

Coleman, Marilyn J., and Lawrence H. Ganong (eds.). 2014. *The Social History of the American Family: An Encyclopedia*. London: Sage.

Cramer, Kathryn. 2003. Hard Science Fiction. In *The Cambridge Companion to Science Fiction*, eds. Edward James and Farah Mendelsohn, 86–96. Cambridge: Cambridge University Press.

James, Edward. 2003. Utopias and Anti-Utopias. In *The Cambridge Companion to Science Fiction*, eds. Edward James and Farah Mendelsohn, 219–229. Cambridge: Cambridge University Press.

Jameson, Fredric. 2005. *Archaeologies of the Future: The Desire Called Utopia and Other Science Fictions*. London: Verso.

Knell, Sebastian, and Marcel Weber (eds.). 2009. *Länger leben? Philosophische und biowissenschaftliche Perspektiven*. Frankfurt am Main: Suhrkamp.

Lem, Stanislaw. 1970. *Solaris*. New York: Walker and Co.

Mangum, Teresa. 2002. Longing for Life Extension: Science Fiction and Late Life. *Journal of Aging and Identity* 7 (2): 69–82.

Mendelsohn, Farah. 2003a. Introduction: Reading Science Fiction. In *The Cambridge Companion to Science Fiction*, eds. Edward James and Farah Mendelsohn, 1–12. Cambridge: Cambridge University Press.

———. 2003b. Religion and Science Fiction. In *The Cambridge Companion to Science Fiction*, eds. Edward James and Farah Mendelsohn, 264–275. Cambridge: Cambridge University Press.

Nye, David. 1984. *The American Technological Sublime*. Cambridge: MIT Press.

Shorter, Edward. 2008. *Before Prozac: The Troubled History of Mood Disorders in Psychiatry*. Oxford: Oxford University Press.

Slonczewski, Joan, and Michael Levy. 2003. Science Fiction and the Life Sciences. In *The Cambridge Companion to Science Fiction*, eds. Edward James and Farah Mendelsohn, 174–185. Cambridge: Cambridge University Press.

Suvin, Darko. 1979. *Metamorphoses of Science Fiction: On the Poetics and History of a Literary Genre*. New Haven: Yale University Press.

Wittenberg, David. 2012. *Time Travel: The Popular Philosophy of Narrative*. New York: Fordham University Press.

Index

© The Editor(s) (if applicable) and The Author(s) 2017
S. Baumbach et al. (eds.), *The Fascination with Unknown Time*,
DOI 10.1007/978-3-319-66438-5

The manufacturer's authorised representative in the EU is Springer
Nature Customer Service Centre GmbH, Europaplatz 3, 69115 Heidelberg,
Germany. If you have any concerns regarding our products, please
contact ProductSafety@springernature.com

Printed and bound by CPI Group (UK) Ltd, Croydon, CR0 4YY

27/04/2026

02097570-0003